620.175.5

CW00370203

PHOTOELASTICITY
IN ENGINEERING PRACTICE

Photoelasticity in Engineering Practice

Edited by

S. A. PAIPETIS

*Professor of Engineering Mechanics,
University of Patras, Patras, Greece*

and

G. S. HOLISTER

*Professor of Engineering Science,
The Open University, Milton Keynes, UK*

ELSEVIER APPLIED SCIENCE PUBLISHERS
LONDON and NEW YORK

ELSEVIER APPLIED SCIENCE PUBLISHERS LTD
Crown House, Linton Road, Barking, Essex IG11 8JU, England

Sole Distributor in the USA and Canada
ELSEVIER SCIENCE PUBLISHING CO., INC.
52 Vanderbilt Avenue, New York, NY 10017, USA

WITH 11 TABLES AND 92 ILLUSTRATIONS

© ELSEVIER APPLIED SCIENCE PUBLISHERS LTD 1985

British Library Cataloguing in Publication Data

Photoelasticity in engineering practice.
1. Photoelasticity
I. Paipetis, Stephen A.   II. Holister, G. S.
620.1'1232   TA418.12

Library of Congress Cataloging in Publication Data

Photoelasticity in engineering practice.

Bibliography: p.
Includes index.
1. Photoelasticity—Addresses, essays, lectures.
I. Paipetis, S. A.   II. Holister, G. S.   III. Title:
Photo-elasticity in engineering practice.
TA418.12.P46   1985   620.1'1295   85–6914

ISBN 0-85334-363-2

The selection and presentation of material and the opinions expressed in this publication
are the sole responsibility of the authors concerned

Printed in Northern Ireland by The Universities Press (Belfast) Ltd

PREFACE

In 1816, Sir David Brewster discovered the phenomenon of temporary double diffraction induced in amorphous transparent materials by mechanical stresses. The next important event, which turned photoelasticity into a powerful stress analysis tool, was the formulation of the Maxwell–Neumann stress-optic law. According to this law, changes in refractive indices are related linearly to the stresses or strains developing in a linearly elastic material and, based on such a relationship, models of structural components can be constructed and analysed so as to determine the state of stress in the model on a whole-field basis.

Out of this fundamental idea, a vast number of experimental techniques were developed, often combined with numerical or analytical procedures, while the introduction of new materials, such as epoxy resins, provided almost unlimited possibilities of modelling practically any mechanical system. The combination with other optical methods, such as moiré, interferometry and holography, along with elaborate photographic techniques produced highly effective procedures for the solution of problems which could not be treated by any other means. Moreover, photoelastic techniques were applied to the investigation of the physical properties of materials under the influence of not only mechanical but also electrostatic, magnetic and electromagnetic fields.

As far as structural problems are concerned, the development of large computers and powerful numerical methods, such as finite-element methods, initially gave the impression that experimental methods and, in particular, photoelastic techniques, were no longer necessary. However, reality turned out to be quite different, since photoelasticity went on developing to encounter new problems, which were either impossible or too costly to be dealt with by numerical means. Recent advances in connection with some of these problems are examined in the present volume, covering such aspects as dynamic photoelasticity, photoplasticity, integrated photoelasticity, applications in fracture mechanics and three-dimensional photoelasticity.

We hope that the latest state of the art in these technologically most important areas, as presented in this book, will be of assistance to modern stress analysts, structural engineers and materials scientists.

S. A. PAIPETIS

G. S. HOLISTER

CONTENTS

Contents

LIST OF CONTRIBUTORS

H. K. ABEN

Institute of Cybernetics, Academy of Sciences of the Estonian SSR, 200104 Tallinn, USSR

I. M. DANIEL

Department of Mechanical Engineering, Illinois Institute of Technology, Chicago, Illinois 60616, USA

E. E. GDOUTOS

Department of Civil Engineering, School of Engineering, Democritus University of Thrace, Xanthi, Greece

JAN JAVORNICKÝ

Institute of Theoretical and Applied Mechanics, Czechoslovak Academy of Sciences, Nové Město, Vyšehradská 49, 128 41 Prague 2, Czechoslovakia

J. I. JOSEPSON

Institute of Cybernetics, Academy of Sciences of the Estonian SSR, 200104 Tallinn, USSR

S. A. PAIPETIS

Department of Mechanical Engineering, University of Patras, Patras, Greece

ix

1

DYNAMIC PHOTOELASTICITY

I. M. Daniel

Illinois Institute of Technology, Chicago, Illinois, USA

ABSTRACT

Dynamic photoelasticity applies the basic photoelastic approach to all areas of elastodynamics, i.e. stress-wave propagation, vibration and impact, quasi-static transient phenomena, and fracture propagation. The most difficult aspect of dynamic photoelasticity is recording the transient isochromatic fringe patterns. Recording methods used to date include high-speed framing cameras ranging in speed from 8000 to 2 000 000 frames per second, the spark-gap (Cranz–Schardin) camera with equivalent rates of 20 000 to 800 000 frames per second, and stop-action single-pulse or strobe systems. Data analysis is based on a dynamic version of the stress-optic law. Stress separation is possible in special cases, such as free or rigid boundaries, axisymmetric problems and pure shear-wave propagation. Viscoelastic response can be studied by using appropriate model materials and photoviscoelastic stress-birefringence relations. Applications are discussed in the areas of: wave propagation in viscoelastic, anisotropic, rock and layered media; water jet cutting; flaw detection; and dynamic fracture.

1. INTRODUCTION

Dynamic photoelasticity applies the basic principles of photoelasticity to the study of high-speed, short-time phenomena. It is applied to all

1

areas of elastodynamics, i.e. stress-wave propagation, vibration, impact, quasi-static transient phenomena, and dynamic fracture. As an optical method, it gives full-field visualisation of transient phenomena and can be applied under a variety of conditions, two- and three-dimensional, elastic and inelastic, isotropic and anisotropic.

As in the case of static photoelasticity, the dynamic counterpart makes use of transparent birefringent materials and modelling techniques. The materials range in stiffness from urethane rubbers to glass. Although direct modelling of dynamic phenomena is not easy, the advantages of dynamic photoelasticity lie in that it gives results difficult to obtain by analytical or other experimental methods. One very useful role of dynamic photoelasticity is the verification of dynamic computer codes.

The first known attempt to apply dynamic photoelasticity was by Tuzi[1] in 1928 using a movie camera at 32 frames per second. Results were improved substantially when Tuzi and Nisida[2] in 1936 used a rotating drum camera at a rate of 1200 frames per second. Additional early contributions were made by Findley,[3] Senior and Wells,[4] Foeppl[5] and Christie.[6] The method received great impetus in the 1950s and 1960s with the use of new materials and the development of improved recording techniques. An early review of dynamic photoelasticity was written by Goldsmith.[7] A more recent review was given by Dally.[8]

The introduction of low-modulus photoelastic materials with wave propagation velocities of the order of $50 \, \mathrm{ms}^{-1}$ allowed the use of conventional high-speed photography. Perkins[9] and Dally *et al.*[10] used a Fastax rotating prism camera to study wave propagation in urethane rubber. The improvement of high speed photographic techniques permitted the use of high-modulus photoelastic materials with wave propagation velocities of the order of $2500 \, \mathrm{ms}^{-1}$. Feder *et al.*[11] and Flynn *et al.*[12] adapted and improved the high-speed framing camera for dynamic photoelasticity. Frocht *et al.*[13] employed streak photography, and Christie[14] and Wells and Post[15] used a multiple-spark (Cranz–Schardin) camera. The application of the Cranz–Schardin camera to dynamic photoelasticity was described by Riley and Dally.[16] Rowlands *et al.*[17] used a sequentially pulsed ruby laser synchronised with a high-speed framing camera. The most recent development in this area is a hybrid Cranz–Schardin laser system by Dally and Sanford[18] which employs ruby lasers with fibre-optic output.

Data analysis and interpretation of results are based on a dynamic version of the stress-optic law. Direct stress separation is possible in special cases, such as free or rigid boundaries, axisymmetric problems, and pure shear-wave propagation.[19] Special techniques, such as the dual-beam polariscope, have been developed for separation of principal stresses in other cases.[20] Stress separation procedures have been developed by Post[21] and Bohler and Schumann[22] by combining photoelastic and interferometric measurements. Stress separation is also obtained by using complementary methods, such as moiré.

Most model materials used in dynamic photoelasticity are viscoelastic and exhibit pronounced time-dependence (or rate-dependence) in their mechanical and optical properties. In most applications the model material is assumed to behave elastically. The usual approximate approach is to characterise the material as a function of strain rate and use the 'dynamic' properties in the elastic stress-optic law. In some cases this approximation is not valid and there is need to account fully for the viscoelastic nature of the material. Dynamic photoviscoelastic methods have been developed and used by Daniel.[23] Dynamic photoelastic methods have been extended to opaque materials using birefringent coatings,[24] to three-dimensional models[25] and anisotropic materials.[26] Most of the earlier applications dealt with wave propagation and dynamic stress concentration problems. The preponderance of more recent work deals with dynamic fracture.[27]

This chapter describes the various aspects of dynamic photoelasticity, including model materials, loading techniques, recording methods, data analysis and interpretation and a variety of applications.

2. FUNDAMENTALS OF DYNAMIC PHOTOELASTICITY

2.1. Elastic Stress-Optic Law

In many applications of the method the material is treated as linear elastic and the same stress-optic relations are carried over from quasistatic elasticity, except that the static material properties are replaced by 'dynamic' ones obtained by appropriate calibration.

The elastic stress-optic law in the general case can be expressed as

$$n_1 - n_0 = C_1\sigma_1 + C_2(\sigma_2 + \sigma_3)$$
$$n_2 - n_0 = C_1\sigma_2 + C_2(\sigma_3 + \sigma_1) \qquad (1)$$
$$n_3 - n_0 = C_1\sigma_3 + C_2(\sigma_1 + \sigma_2)$$

where: C_1, C_2 = stress-optic coefficients; n_1, n_2, n_3 = principal indices of refraction; and n_0 = initial index of refraction in unstressed isotropic body. Because of the difficulties involved in measuring the principal indices of refraction and the principal optical directions, photoelasticity is usually confined to measuring relative birefringence.

$$N = (n_1 - n_2)\frac{h}{\lambda} = (C_1 - C_2)(\sigma_1 - \sigma_2)\frac{h}{\lambda}$$

$$= \frac{h}{f_\sigma^*}\frac{\sigma_1 - \sigma_2}{2} \tag{2}$$

or

$$\sigma_1 - \sigma_2 = \frac{2f_\sigma^*}{h}N \tag{3}$$

where: N = fringe order; h = length of optical path and, in some cases, thickness of specimen; σ_1, σ_2 = principal stresses acting on planes parallel to axis of light propagation; and f_σ^* = dynamic material fringe value for a given wavelength λ.

In addition to isochromatic fringes, one can obtain isoclinic fringes, i.e. loci of points of constant angle ϕ, between the principal optical axis and a reference x-axis. If stresses σ_1 and σ_2 lie on the x–y plane, we have the following expressions for the cartesian components of stress

$$\sigma_x - \sigma_y = \frac{2Nf_\sigma^*}{h}\cos 2\phi_{1-x}$$

$$\tau_{xy} = -\frac{Nf_\sigma^*}{h}\sin 2\phi_{1-x} \tag{4}$$

where ϕ_{1-x} = angle between x-axis and maximum secondary principal stress σ_1 measured positive counterclockwise.

These relationships are useful in determining the complete state of stress using some other auxiliary method or complementary data, such as strains from moiré patterns.

2.2. Viscoelastic Stress-Optic Law

The time-dependence of mechanical and optical properties of model materials can best be described in terms of linear viscoelasticity. In the usual treatment of the linear theory of viscoelasticity it is customary to describe material behaviour separately under pure shear and pure dilatation. Then, the constitutive relations can be written in the form

$$s_{ij} = 2 \int_0^t G(t-\tau) \frac{de_{ij}(\tau)}{d\tau} d\tau$$

$$\sigma = 3 \int_0^t K(t-\tau) \frac{d\varepsilon(\tau)}{d\tau} d\tau \tag{5}$$

where $G(t)$ and $K(t)$ are the shear and bulk relaxation moduli. Several forms of the stress-optic law for viscoelastic materials have been proposed.[23,28-34] One form convenient for experimental applications is the integral form which is analogous to the constitutive relations of eqn. (5). The normal stress difference and the shear stress are given by

$$\sigma_{xx} - \sigma_{yy} = \frac{2}{h} \int_0^t f_\sigma(t-\tau) \frac{d}{d\tau} [(n_1 - n_2) \cos 2\phi_n] d\tau \tag{6}$$

$$\sigma_{xy} = \frac{1}{h} \int_0^t f_\sigma(t-\tau) \frac{d}{d\tau} [(n_1 - n_2) \sin 2\phi_n] d\tau \tag{7}$$

where: $n_1 - n_2 =$ relative birefringence (or difference of secondary principal indices of refraction); $\phi_n =$ angle between principal optical direction and x-axis; and $f_\sigma(t) =$ time-dependent stress fringe value.

2.3. Material Characterisation

A variety of model materials have been used for dynamic photoelastic studies. Transparent urethane rubber has been used for wave propagation studies using moderate speed cameras. Plasticised polyvinyl chloride (PVC) has been used for photoviscoelastic studies. Harder materials used are Columbia resin (CR-39), epoxies and polyesters. One of the latter, Homalite 100, is commonly used in dynamic fracture studies. A number of investigators, including Dally *et al.*,[10] Williams and Arenz,[31] Khesin *et al.*,[32] Clark,[35] Clark and Sanford,[36] Frocht,[37] Daniel,[38] Brown and Selway,[39] Chase and Goldsmith[40] and Peeters and Parmerter,[41] have studied the mechanical and optical properties of a number of photoelastic materials under dynamic conditions.

For the elastic approximation, material characterisation is accomplished by impacting the end of a prismatic bar specimen and measuring the force, deformation and birefringence in the specimen as functions of time. For soft urethane rubber Dally *et al.*[10] used a short prismatic specimen impacted between two pendulums, one initially stationary and one moving. The specimen is loaded in dynamic compression and the acceleration imparted to the stationary pendulum is measured. Simultaneously, the birefringence induced in the specimen is measured either by means of a high-speed framing camera or a photodetector. Results were expressed as functions of strain rate. The modulus increases from a static value of 3060 kPa to a value of 5550 kPa at a strain rate of 63 s^{-1}. The stress fringe value increases by approximately 10% over this strain rate range. Similar results were obtained by using a strut under axial impact.

Clark and Sanford calibrated several epoxies and Homalite 100 by using a long bar of the material with square cross-section impacted at one end by a projectile fired from an air gun.[36] Strain gauges were mounted on two opposite sides at the centre of the bar, and polarised light was transmitted across the other two sides of the bar at the same location. The variation in light intensity, related to birefringence, $N(t)$, was monitored with a photomultiplier tube. A second plastic bar (throw-off bar), abutting the back end of the specimen, collects the momentum. The dynamic modulus is obtained as

$$E_d = \frac{mv}{A \int_0^T \varepsilon(t)\, dt} \tag{8}$$

where: A = cross-sectional area of specimen; $\int_0^T \varepsilon(t)\, dt$ = area under the strain versus time curve; m = mass of throw-off bar; and v = velocity of throw-off bar.

In an alternative approach, a second set of strain gauges is mounted at another station in the bar to measure the wave propagation velocity. One set of gauges consists of two-gauge rosettes so that Poisson's ratio can be obtained as

$$\nu = -\frac{\varepsilon_t(t)}{\varepsilon_a(t)} \tag{9}$$

where $\varepsilon_t(t)$ and $\varepsilon_a(t)$ are transverse and axial strains, respectively. The dynamic modulus, strain fringe value and stress fringe value are then obtained as follows:

$$E_d = c_0^2 \rho \tag{10}$$

$$f_\varepsilon^* = \frac{h(1+\nu)}{N(t)} \varepsilon_a(t) \tag{11}$$

$$f_\sigma^* = \frac{E_d}{1+\nu} f_\varepsilon^* = \frac{hc_0^2\rho}{N(t)} \varepsilon_a(t) \tag{12}$$

where c_0 = wave propagation velocity, ρ = density of specimen material, h = specimen thickness (light path) and f_σ^* and f_ε^* = dynamic stress and strain fringe values, respectively.

For the epoxies and the Homalite 100 tested, with a loading time of approximately 200 μs, the dynamic modulus and stress fringe value were approximately 10–15% higher than the static values.

Since all polymeric model materials are viscoelastic to a lesser or greater extent, their dynamic properties are best represented as viscoelastic functions of time or frequency. Preferred forms of these properties are the relaxation modulus and the stress fringe value, both functions of time. However, it is more convenient to determine them experimentally as functions of frequency.

Short-time (dynamic) behaviour is usually measured experimentally by means of sinusoidal oscillation tests. For sinusoidal variation of stress, the strain will also vary sinusoidally, although not in phase with the stress variation. Thus, if

$$s_{ij} = s_0 \sin \omega t \tag{13}$$

then

$$e_{ij} = e_0 \sin (\omega t - \delta) \tag{14}$$

where ω = angular frequency of oscillation and δ = phase difference.

The stress ratio, called complex modulus, is given by

$$G^*(i\omega) = \frac{s_{ij}}{2e_{ij}} = G_1(\omega) + iG_2(\omega) \tag{15}$$

where $G_1(\omega)$ and $G_2(\omega)$ are the real and imaginary components of the complex modulus, respectively.

The complex modulus can be converted into a relaxation modulus for use in the constitutive relation (eqn (5)), by means of exact or approximate relationships. Exact relationships are

$$G(t) = G_e + \frac{2}{\pi} \int_0^\infty \frac{G_1 - G_e}{\omega} \sin \omega t \, d\omega \tag{16}$$

or

$$G(t) = G_e + \frac{2}{\pi} \int_0^\infty \frac{G_2}{\omega} \cos \omega t \, d\omega \tag{17}$$

where

$$G_e = \lim_{t \to \infty} G(t) = \text{equilibrium modulus}$$

An approximate expression is given by Ninomiya and Ferry:[42]

$$G(t) \simeq G_1(\omega) - 0 \cdot 40 G_2(0 \cdot 4\omega) + 0 \cdot 014 G_2(10\omega)|_{\omega = 1/t} \tag{18}$$

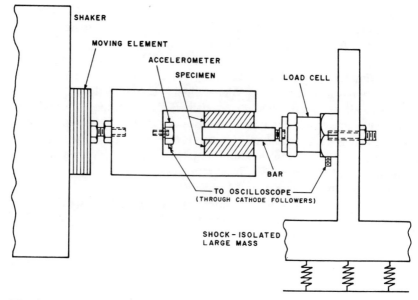

Fig. 1. Schematic diagram of setup used for determination of complex shear modulus.[38]

For materials with moderate loss ($\tan \delta \ll 1$)

$$G(t) \simeq G_1(\omega)\big|_{\omega = 1/t} \tag{19}$$

Viscoelastic properties as a function of frequency were determined by Daniel[38] for plasticised polyvinyl chloride. The complex shear modulus was obtained by loading two prismatic specimens under sinusoidal oscillations in the fixture shown in Fig. 1. The specimens ($3 \cdot 3 \times 0 \cdot 64 \times 1 \cdot 0$ cm) were cemented to a metal fork and a central bar so that motion of the fork with respect to the bar subjects them to simple shear. The central bar was connected through a load cell to a large shock-isolated floating mass, which can be regarded as rigid for all practical purposes. The fork was connected to the moving end of a shaker producing sinusoidal oscillations.

The shear strain in the specimen was obtained from the acceleration of the fork. The shear stress was determined from the transmitted force, which was measured with a load cell. The signals from these two transducers were recorded simultaneously on an oscilloscope for various frequencies. From the amplitude and phase relationship of these

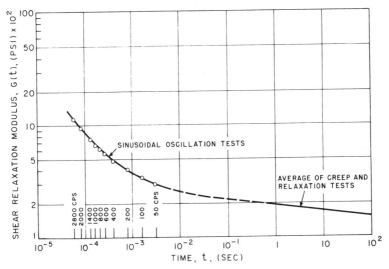

Fig. 2. Shear relaxation modulus of polyvinyl chloride as a function of time.[38]

two signals, the real and imaginary components of the modulus were determined. Finally, the relaxation shear modulus was computed using the approximate relation (eqn (18)) and plotted in Fig. 2 as a function of time.

In the sinusoidal oscillation test, stress and birefringence may in general be out of phase. Then, a complex stress fringe value as a function of frequency can be defined as

$$f_\sigma^*(i\omega) = f_{\sigma 1}(\omega) + i f_{\sigma 2}(\omega) \tag{20}$$

This function of frequency can be converted into a function of time by means of exact or approximate interrelations for use in the stress-optic expressions, eqns (6) and (7). The dynamic fringe value was determined by subjecting a prismatic specimen to sinusoidal uniaxial stress of various frequencies in the fixture and set-up shown in Fig. 3. The applied stress was obtained from the load cell signal. The transmitted polarised light was measured by a photocell. When the system is set for maximum sensitivity, the fringe order is obtained as[38]

$$N = \frac{1}{2\pi} \arcsin \left[\frac{2\Delta I}{I_0} \right] \tag{21}$$

where ΔI = change in light intensity and I_0 = maximum intensity of

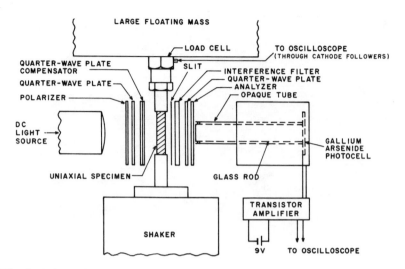

Fig. 3. Schematic diagram of setup used for determination of dynamic fringe value.[38]

light transmitted through elements of circular polariscope, compensator, and model.

The stress fringe value, obtained as a function of frequency, was converted to one as a function of time (relaxation type) by an approximate relation similar to eqn (18). The result is shown in Fig. 4.

A wave propagation technique was developed by Peeters and Parmerter[41] for determining the dynamic stress fringe value directly as a function of time on a microsecond timescale. The technique is analogous to that developed by Sackman and Kaya[43] for determination of short-time viscoelastic properties. The basic procedure consists of measuring particle velocity at two stations along a rod along which propagates a longitudinal pulse. In general, the pulse is dispersed and attenuated between the two stations, and the change in pulse shape is related to the relaxation function. The method was demonstrated for an Araldite epoxy in the time range 1–100 μs. Results were expressed in the form of an inverse optical creep function of time.

The case of very large strains at high strain rates, where linear viscoelasticity is not valid, was treated by Chase and Goldsmith.[40] They proposed two empirical constitutive relations to characterise the uniaxial photomechanical behaviour of a polyester–styrene copolymer for strain rates up to $3000\,s^{-1}$ and strains up to 40%.

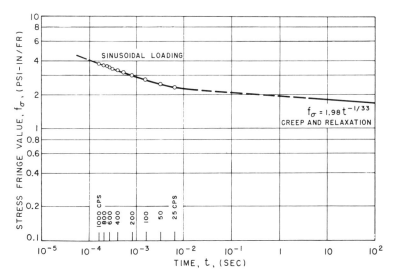

Fig. 4. Stress fringe value of polyvinyl chloride as a function of time.[38]

3. RECORDING METHODS

One of the most important aspects of dynamic photoelasticity is recording the transient isochromatic fringe patterns. Depending on the model material and type of loading, these fringes propagate at velocities between 50 and 2500 ms^{-1}. Several types of recording system are possible. They include single flash photography, streak cameras, framing cameras, multiple spark-gap (Cranz–Schardin) systems, laser systems, stop action or strobe systems and digital imaging cameras. Discussions of some of the recording methods have been given by Dally[8] and Clark and Durelli.[44]

3.1. Single Flash Photography
If the dynamic loading on the model can be repeated precisely and the specimen response is reproducible, it is possible to use a single spark-gap light source and an inexpensive camera. A typical flash unit (EG & G Microflash System) consists of an air flashtube, a reflector, the rectifier transformer and a power supply. A 2 kV pulse triggers the flashtube and generates a brilliant flash (50×10^6 beam candlepower) of 0–5 μs duration. The flash unit can be used with a delay generator

for synchronisation with the various stages of an event. Single flashes of high-intensity light of short duration (40 ns) can be produced with a modulated ruby laser.[45] Ultrashort single and repeated light pulses can be produced by equipment such as the Hamamatsu Picosec Light Pulser.

Single flash photography is used in conjunction with repeatable and controlled loading such as hinged swinging hammer,[46] falling weight,[47] air shock loading,[48] airgun driven projectiles, explosives, electromagnetic transducers and water jets. It allows the use of large size film and gives good definition of details, especially around stress concentrations. It cannot be used when non-linear deformation or fracture take place during the test.

3.2. Framing Cameras

There is a wide range of medium to high-speed framing cameras which can be used in dynamic photoelastic studies with conventional high-intensity light sources. One type of such cameras makes use of a continuously moving roll of film with a rotating optical system (mirror or prism) which holds the image formed by the main lens stationary with respect to the moving film. Such cameras include the Fastax, Hycam (Redlake) cameras and use 100 ft rolls of 16 mm or 35 mm film. They can record events at rates varying from a few hundred to over 20 000 frames per second, with exposure times typically equal to one-third the interframe time interval. A common light source for these cameras is a number of flashbulbs with light duration of up to 100 ms. These medium speed framing cameras have been used extensively in photoelastic studies employing 'soft' model materials, such as urethane rubber and plasticised PVC, with propagation velocities of the order of $50 \, \text{ms}^{-1}$.

Higher modulus materials require higher framing rates. This is usually accomplished by combining a rotating mirror with a stationary strip of film on a drum. The light from the model is focused by the main lens on a rotating mirror and projected and focused onto the film through pairs of slits and relay lenses. With the normal aperture slits used the exposure time on each frame is approximately one-third the interframe time interval. The first application of a high-speed framing camera, such as the Beckman and Whitley, was reported by Feder *et al.*[11] in 1956. Subsequently Flynn *et al.*[12,49] introduced many improvements in the application of this camera to a variety of problems in dynamic photoelasticity. The Beckman and Whitley camera is typical of the sort used in many photoelastic applications. This camera is capable of recording 25 35 mm frames at up to 1 250 000 frames per second. Similar cameras with framing rates of up to 10 000 000 frames per second and 50–100 frame capability are also available.[8,44] In order to achieve exposure times of less than 1 μs it is necessary to keep the framing rate above 300 000 frames per second.

High-speed framing cameras require high-intensity illumination because of the short exposure times and the losses due to the polariscope elements. Two types of continuous light source have been used, argon flashbombs or an electronic flash. The argon flashbomb is a clear latex balloon inflated with argon gas. The gas is ignited with a detonator. The light produced has a rise time of approximately 12 μs, with the illumination relatively constant for 40 μs. The destructive nature of the argon bomb detonation makes it suitable primarily in outside field experiments. The electronic light source employs one or more flash lamps (xenon) which are energised and switched by a specially designed power supply and circuit.[26] An oscilloscope record of the light output from such an electronic flash is shown in Fig. 5. This flash has a rise time of approximately 45 μs and a controlled effective duration varying from 20 to 100 μs. The illumination is comparable to that produced by one argon bomb.

A wide variety of films and developers with varying processing times have been employed. Combinations used include high-speed Ektachrome with white light, 4-X film processed in D-76 or Acufine developer, Tri-X panchromatic processed in Edwal FG-7, Double-X

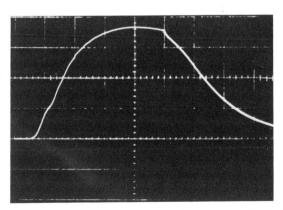

Fig. 5. Oscilloscope record of light output from electronic flash (FT 220 flashlamp; input energy 200 W-s at 2 kV, 100 μfd; sweep = 20 μs cm^{-1}).[26]

film processed in Acufine, Dektol or D-76 developers, and Kodak 2485 processed in Kodak developer 857. Best results have been obtained with Double-X film developed in either Dektol or D-76.[26,49]

Framing cameras with nanosecond resolution and effective rates of 10^8 frames per second have been developed, but to the best of our knowledge have not been applied to dynamic photoelasticity.[50]

3.3. Streak Photography

Many framing cameras can be adapted for this mode of operation. A streak photograph, in contrast to a full image photograph, is obtained from a very narrow rectangular zone of the model. The model is normally masked so that light passes only through a slit in the mask. The slit of light is received by the camera and focused on a continuously moving film. The resulting photograph gives a continuous time record of the birefringence along the line being photographed. The effective time of exposure is calculated as

$$t = \frac{d}{v} \tag{22}$$

where d = width of slit image on the film and v = linear velocity of the film. For purposes of comparison with framing photography the equivalent framing rate is given by

$$R = \frac{1}{t} = \frac{v}{d} \tag{23}$$

The application of streak photography to dynamic photoelasticity was described by Frocht *et al.*[13] They achieved exposure times of $\frac{2}{3}$ μs and obtained fringe patterns in discs and beams under impact.

3.4. Multiple Spark Systems

The multiple spark gap camera, originally developed by Cranz and Schardin[51] in 1929, has been applied by many investigators. Christie[14] described the first application to photoelasticity; Wells and Post[15] applied it to the problem of a running crack; and Riley and Dally[16] employed the system in a large number of geophysical problems. The Cranz–Schardin system consists of three subsystems: the spark-gap assembly, the optical bench and the control and synchronisation circuits.

The spark-gap assembly consists of an array of spark gaps which are fired in sequence to provide controlled illumination. The spark gaps

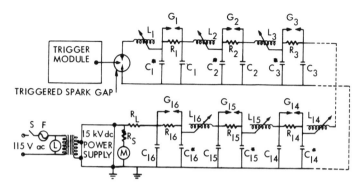

Fig. 6. LC delay loop circuit energising an array of spark gaps.[16]

are connected in an LC delay loop circuit (Fig. 6). The firing sequence is initiated by applying a 20 kV voltage to the trigger gap. This initiates a sequence of discharges across the spark gaps. The optical subsystem illustrated in Fig. 7 performs three distinct functions, polarisation, image separation and magnification. The light is circularly polarised by surrounding the specimen with two circular polarisers (HNCP-38). The light from each spark gap passes through the first polariser, the transparent specimen, the second polariser (analyser), and then it projects and focuses the image of the specimen on a separate camera lens. The images are separated by the geometric positioning of the spark gaps and camera lenses. For a typical 4×4 array of spark gaps, 16 distinct images, corresponding to 16 different time instances in the event, are obtained on one sheet of film. The time sequence of the

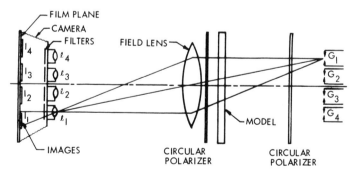

Fig. 7. Arrangement of elements in optical subsystem of Cranz–Schardin camera.[16]

sparking is controlled by the inductances in the control circuit.[16] The magnification is controlled by the lenses selected. A typical magnification ratio is -0.1.

The light pulse duration of a spark gap is approximately $0.5\ \mu s$ which is adequate for recording fast-moving densely spaced fringes (0.6 fringes mm^{-1} at $2500\ ms^{-1}$).[8] The light intensity is high enough to record images in relatively slow (ASA 16) film, after filtering with blue ($440\ \mu m$) wavelength filters. The interframe time is adjustable and the effective framing rate can be varied between 20 000 and 800 000 frames per second.

The loading of the specimen, the initiation of spark-gap firing, and recording the time of each frame are controlled and synchronised to the nearest microsecond with a circuit such as the one shown in Fig. 8.

In later versions of the camera a second field lens was added between the light sources and the first polariser. The spark-gap assembly was enclosed in a remote location and the light from each spark gap was channelled through a fibre optic bundle to a board on the optical bench. A simplified version of the Cranz–Schardin camera was proposed by Conway.[52] It consists of trigger and timing circuits, pulse generator circuits (one for each frame) and flash bulb circuits (one for each frame). The spark gaps are replaced by xenon flash bulbs which require $+400$ V and -400 V across their plates to discharge the tube.

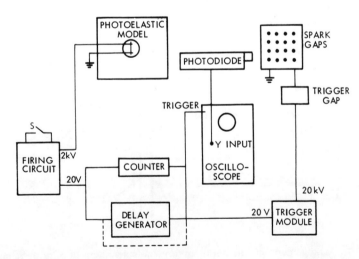

Fig. 8. Synchronisation and control circuits for a Cranz–Schardin camera.[16]

In the latest development reported by Dally and Sanford[18] the spark-gap light sources have been replaced by ruby lasers with fibre optic light guides. Six independent Q-switched ruby lasers provide six very intense and ultra-short (30 ns) pulses of light. A microprocessor provides precise synchronisation by computing in real time the anticipated arrival time and controlling the Q-switching of the lasers. The advantages of this system are:

1. Very short exposure times (30 ns).
2. Extremely high framing rates (up to 10^7 frames per second).
3. High intensity light output.
4. Improved control and synchronisation.
5. Versatility and adaptability to various experiments.
6. Highly monochromatic light.

The main disadvantage of this system is its high cost.

3.5. Laser Systems

The laser, especially the pulsed ruby laser, provides an ideal light source for dynamic photoelasticity. The output of a ruby laser is a highly collimated beam of monochromatic, plane-polarised light of high intensity and extremely short (30 ns) duration. This output can be modulated by means of a modulator, such as a Pockels cell, to obtain a single pulse or a series of short pulses.[17] The Pockels cell modulator behaves like an electrically activated quarter-wave retardation plate, which rotates the plane of polarisation of the laser light in proportion to the voltage applied to the Pockels cell. By controlling the electrical pulses across the modulator, one can produce a series of short light pulses (Q-switched pulses). The ruby laser has been Q-switched at above 170 000 pulses per second.

Initially the laser was used in the single-pulse mode to produce single frames in a repeatable experiment. Rowlands *et al.*[17] successfully coupled a sequentially pulsed ruby laser with a high-speed framing camera. A schematic diagram of the system is shown in Fig. 9. The streak capability of a Beckman and Whitley camera with the slit removed was used. The light entering the camera is collected by the objective lens, reflected by the rotating mirror, and projected onto the stationary film. For a continuous light source the rotation of the mirror simply smears the light beam (image) along the film. Modulation of the laser into a series of short light pulses eliminates smearing and produces a series of separated images, i.e. a framing capability is produced.

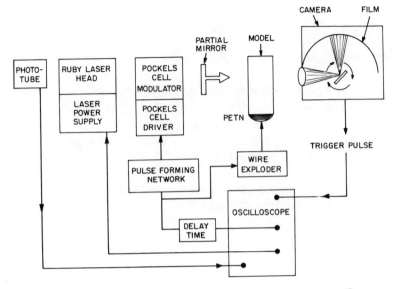

Fig. 9. Block diagram of multiple-pulse ruby laser system.[17]

The system above was applied to transmitted- and scattered-light photoelasticity. Synchronisation of events is accomplished by means of a trigger pulse from the camera which discharges the laser and triggers the explosive loading of the model.

Hendly *et al.*[53] have also developed a hybrid Cranz–Schardin laser system, where the beam from a single sequentially pulsed laser is deflected by an acousto-optic deflector to produce up to five images.

3.6. Strobe Systems

This type of recording method is suitable for dynamic problems which are precisely reproducible and repeatable at a known fixed frequency. A simple method was introduced by Becker[54] for the study of impact problems. Impact loading was repeated at constant frequency by means of a solenoid, and a strobe light was flashed at a controlled frequency close to that of the loading.

When the strobe light and impact loading have the same frequency the image appears stationary and can be photographed easily. Whenever there is a slight mismatch between the two frequencies the fringes appear to move slowly in proportion to the frequency difference. Thus, it is possible to observe and record the event at different stages. The duration of the strobe flash is not relevant, since synchron-isation of the flashing and loading reduces the fringe velocity to nearly zero. In fact, the strobe light can flash several times during the exposure time. Strobe lights are available with flashing frequencies up to 6000 per second.

The method has also been applied by Tschinke[55] to the study of transient stresses around blade roots of rotating turbine discs. Using a low modulus material for the model and high speed stroboscopic techniques he photographed changes in the isochromatic fringe pat-terns as the model was accelerated from 0 to 3000 rpm with a uniform angular acceleration.

3.7. Electronic Methods

Photodetectors, in the form of photocells, photomultipliers or photo-diodes, have been used for measuring dynamic birefringence for a long time. Since they measure birefringence only at a point, they are used primarily in material characterisation studies. Most of the investiga-tions on optical characterisation mentioned before made use of such photosensitive devices.[31,32,35,36,38–41,56] Similar recording techniques have been applied by Schwieger[57] and Cunningham *et al.*[58] for impact and wave propagation problems.

The most recent development in this area is that of the digital imaging camera. A new field known as 'half-fringe photoelasticity' has emerged.[59] The technique is based on a digital grey level analyser and picture processing system. The intensity field of a half-fringe light-field photoelastic pattern can be divided into 256 levels from 'light' to 'dark' for an equivalent fringe multiplication of 512 times. This method is a full-field one and would allow the use of very low intensity loads and/or materials of low birefringence sensitivity, such as glass, trans-parent composites, transparent ceramics, etc. Applications of the method to dynamic photoelastic problems are under study.

4. DATA ANALYSIS

4.1. Introduction

In general, the only data recorded in a two-dimensional dynamic photoelastic experiment are the isochromatic fringe patterns. In an elastic material the fringe order is directly related to the difference of principal stresses by the stress-optic law (eqn (3)) discussed before.

Additional information is needed to obtain the complete state of stress. If the experiment is precisely reproducible, dynamic isoclinic fringes can be obtained by resetting the polariscope and repeating the test for each isoclinic parameter. This would allow the determination of the Cartesian shear stress component in addition to the principal stress difference. Additional information is still needed for the complete separation of stresses.

Dally[19] discussed methods of stress separation for specific cases of special boundary conditions or special types of stress waves, using only isochromatic and isoclinic data. Special boundary conditions include free and rigid boundaries. Special types of waves include axisymmetric radial, plane dilatational, plane shear, spherical dilatational and spherical shear waves.

4.2. Special Boundary Conditions

For any combination of waves propagating along a free boundary the only non-zero principal stress tangential to the boundary is given by

$$\sigma_1 = 2Nf_\sigma^*/h \tag{24}$$

The displacement u along the boundary (x-axis) can also be obtained by using elastic stress–strain relations and strain-displacement relations:

$$u = \frac{2f_\sigma^*}{Eh} \int_B N \, dx \tag{25}$$

For a plane dilatational wave propagating along a rigid boundary (x-axis) the conditions of zero displacement and rotation imply

$$\varepsilon_{xx} = \gamma_{xy} = 0, \qquad \varepsilon_{yy} = \frac{\partial v}{\partial y} \tag{26}$$

where v = displacement component along y-axis normal to the boundary. Therefore, the x- and y-axes are principal directions and

$$\sigma_1 = \frac{2}{1-v} (Nf_\sigma^*/h)$$

$$\sigma_2 = \frac{-2v}{1-v} (Nf_\sigma^*/h) \tag{27}$$

4.3. Special Types of Stress Waves

A point source in the interior of a large plate produces an axisymmetric wave with a displacement field given by

$$
\begin{aligned}
u_r &= f(r) \\
u_\theta &= 0
\end{aligned}
\tag{28}
$$

where u_r, u_θ = radial and circumferential displacements, respectively, and r = radial distance from source. The special strain-displacement relations in this case, the stress–strain relations, and the stress-optic law yield:

$$
\begin{aligned}
\sigma_{rr} &= \frac{E}{1 - \nu^2} (\varepsilon_{rr} + \nu \varepsilon_{\theta\theta}) \\
\sigma_{\theta\theta} &= \frac{E}{1 - \nu^2} (\varepsilon_{\theta\theta} + \nu \varepsilon_{rr})
\end{aligned}
\tag{29}
$$

where the radial and circumferential strains can be obtained from the isochromatic data as follows:

$$
\begin{aligned}
\varepsilon_{rr} &= -\frac{2(1+\nu)}{E} \cdot \frac{f_\sigma^*}{h} \left(\int_r \frac{N}{r} \, dr + N \right) \\
\varepsilon_{\theta\theta} &= -\frac{2(1+\nu)}{E} \cdot \frac{f_\sigma^*}{h} \int_r \frac{N}{r} \, dr
\end{aligned}
\tag{30}
$$

In the case of a plane dilatational wave propagating along the x-axis, the conditions of zero rotation imply that the x-axis is a principal axis and a condition of plane strain exists with $\varepsilon_{yy} = 0$. Therefore,

$$
\sigma_{yy} = \nu \sigma_{xx}
\tag{31}
$$

and, from the stress-optic law,

$$
\begin{aligned}
\sigma_{xx} &= \frac{2}{1 - \nu} \frac{f_\sigma^*}{h} N \\
\sigma_{yy} &= \frac{2\nu}{1 - \nu} \frac{f_\sigma^*}{h} N
\end{aligned}
\tag{32}
$$

For a plane shear wave it is known that

$$
\sigma_1 = -\sigma_2
\tag{33}
$$

Therefore,

$$\sigma_1 = -\sigma_2 = \frac{Nf_\sigma^*}{h} \qquad (34)$$

For a general two-dimensional dilatational or shear wave, the displacement components can be obtained in terms of isochromatic and isoclinic data and, subsequently, the stress components are obtained by using stress-displacement relations. For an outgoing spherical dilatational or shear wave, the radial displacement, and hence the radial and circumferential strains and stresses, can be obtained in terms of isochromatic data only. For a shear wave propagating along the axis of a cylinder, the only non-zero circumferential displacement component can be obtained in terms of isochromatic data only. Subsequently, strain and stress components are also obtained in terms of isochromatic data only.

4.4. Auxiliary Methods
For points lying on an axis of stress symmetry, separation of principal stresses can be obtained by combining photoelastic data from normal and oblique incidence. For example, when an oblique incidence pattern is obtained by a rotation θ about the σ_2-direction (axis of symmetry), the fringe order N_θ is

$$N_\theta = (\sigma_1 \cos^2 \theta - \sigma_2) \frac{h}{f_\sigma^* \cos \theta} \qquad (35)$$

This relation, when combined with the normal incidence stress-optic law

$$N = (\sigma_1 - \sigma_2) \frac{h}{f_\sigma^*} \qquad (36)$$

yields the individual values for the principal stresses

$$\sigma_1 = \frac{f_\sigma^*}{h \sin^2 \theta} (N - N_\theta \cos \theta) \qquad (37)$$

$$\sigma_2 = \frac{f_\sigma^*}{h \sin^2 \theta} (N \cos^2 \theta - N_\theta \cos \theta) \qquad (38)$$

These equations involve differences between quantities of the same order of magnitude, so that small errors in the measurement of N and N_θ may lead to larger errors in the determination of the stress. Flynn[20] developed a system combining a dual-beam polariscope and a high-speed framing camera to obtain, simultaneously, normal and oblique incidence dynamic fringe patterns.

4.5. Complementary Methods

Complementary non-photoelastic methods can be used to provide additional data for the separation of principal stresses. Durelli *et al.*[60] combined the embedded grid method and dynamic photoelasticity to determine individual stress values in dynamically loaded urethane rubber models.

The moiré method of measuring dynamic surface strains has been used extensively in conjunction with dynamic photoelasticity. Riley and Durelli[61] used moiré rulings of 40 lines mm^{-1} on urethane rubber models to obtain complementary strain data. Daniel[23,62] used similar moiré rulings on soft photoviscoelastic models (PVC), as well as on hard Columbia resin (CR-39) specimens.

In both methods mentioned above two Cartesian strain components, ε_x and ε_y, are obtained by the complementary method. Then,

$$\sigma_x + \sigma_y = \sigma_1 + \sigma_2 = \frac{E}{1-\nu}(\varepsilon_x + \varepsilon_y) \qquad (39)$$

which, when combined with the stress-optic relation (eqn (35)), gives

$$\sigma_1 = \frac{E}{2(1-\nu)}(\varepsilon_x + \varepsilon_y) + \frac{Nf_\sigma^*}{h} \qquad (40)$$

$$\sigma_2 = \frac{E}{2(1-\nu)}(\varepsilon_x + \varepsilon_y) - \frac{Nf_\sigma^*}{h} \qquad (41)$$

Interferometric methods in general, and holographic interferometric methods in particular, have been used to provide the complementary information needed for separation of principal stresses. Interferometry gives information on the thickness changes of the specimen which, under conditions of plane stress, are proportional to the sum of the normal stresses. It is possible with interferometry to obtain loci of points of constant thickness, or isopachics, which are also the loci of points of constant sum of normal stresses. When the isopachics are combined with the isochromatics (loci of points of constant difference of principal stresses) a complete separation of stresses is possible. This approach was originally recommended by Favre[63] and extended and applied by many workers including Post,[64] Nisida and Saito,[65] Fourney,[66] Hovanesian *et al.*,[67] Clark and Durelli,[68] Sanford and Durelli,[69] Holloway *et al.*,[70] Dudderar and Doerries[71] and Lallemand and Lagarde.[72]

In holographic interferometry of a loaded transparent model the light intensity is a function of both the sum and the difference of the principal stresses:

$$I = \frac{a^2}{2} \left\{ 1 + \cos \left[\frac{2\pi}{\lambda} \frac{A' + B'}{2} (\sigma_1 + \sigma_2) h \right] \right.$$
$$\left. \times \cos \left[\frac{2\pi}{\lambda} \frac{C}{2} (\sigma_1 - \sigma_2) h \right] \right\} \tag{42}$$

where A', B' and C are stress-optic constants. The intensity equation above represents two sets of fringes, isochromatics and isopachics. These fringes normally interact in a complicated way and are difficult to interpret. Considerable effort has gone into elimination of the interaction effects and simultaneous recording of two distinct sets of fringes.[70–72]

The complete separation of stresses requires three pieces of information, the individual values of the principal stresses as well as the principal directions. The interferometric methods discussed before give full field information on the sum and difference of principal stresses (isopachics and isochromatics) but not the principal directions. A point-per-point method for the complete determination of a dynamic state of stress was presented by Bohler and Schumann.[22] This is accomplished by making three simultaneous measurements of light intensity of three beams by means of photomultipliers. One beam is a transmitted circularly polarised beam whose intensity is related to the relative retardation (difference in principal stresses); a second beam is a transmitted plane-polarised one whose intensity is related to the relative retardation and the principal directions (isoclinic); and a third beam, obtained by multiple reflections from the two faces of the specimen, and whose intensity is related to both the mean absolute retardation (sum of principal stresses) and relative retardation (difference of principal stresses). The individual principal stresses and the isoclinic angle can be expressed in terms of three light intensity measurements.

5. APPLICATIONS

5.1. Introduction

Dynamic photoelasticity has been applied to all areas of elasto-dynamics, such as wave propagation, vibration, fracture mechanics, jet cutting and flaw detection. Examples of various applications will be described below.

5.2. Elastic Wave Propagation

The problem of wave propagation is of particular interest in seismology, explosive mining and excavation, and design of blast-resistant structures. The problem of wave propagation in isotropic elastic media has been dealt with extensively by analytical and experimental methods. Dynamic loading in an infinite medium generates two types of waves, a dilatational wave propagating with a velocity of

$$c_1 = \sqrt{\frac{\lambda + 2\mu}{\rho}} \tag{43}$$

and a distortional wave propagating with a velocity of

$$c_2 = \sqrt{\frac{\mu}{\rho}} \tag{44}$$

where c_1 and c_2 = dilatational and distortional wave propagation velocities, respectively, λ, μ = Lamé elastic constants and ρ = density.

In addition to these two basic waves a third type occurs along free surfaces. This wave, referred to as a Rayleigh wave, travels with a velocity slightly lower than that of the distortional wave.

Early photoelastic studies of wave propagation made use of low modulus model materials loaded by falling weights, explosive charges, or air shock.[4,9–11,48] Cylindrical or plane stress waves and their interaction with geometric discontinuities, holes and inclusions, were studied. Dynamic photoelasticity was complemented with the moiré method of strain analysis. Dynamic stress distributions around the boundaries of holes were obtained and compared with equivalent 'static' stress distributions. In the case of air shock loading the experimental results were in good agreement with theoretical predictions.[48] Results indicated that the dynamic stress-concentration factor is initially lower than the static one, exceeds it at a later time, and finally approaches it asymptotically.

Model materials for dynamic photoelastic studies can be screened and evaluated by means of simple wave propagation tests. These tests are conducted by loading explosively 6·4 mm thick plates of dimensions 10×25 cm, and recording the transient fringe patterns. Figure 10 shows typical fringe patterns in a Columbia resin (CR-39) specimen, obtained with a spark-gap camera operating at an effective rate of

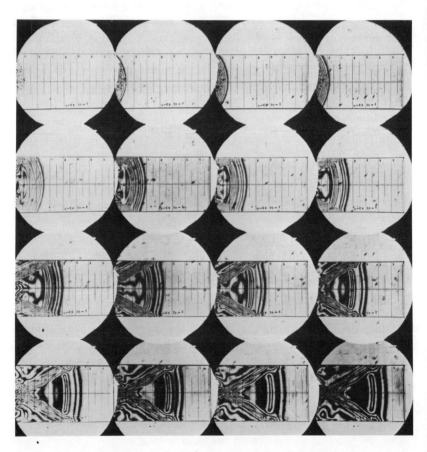

Fig. 10. Isochromatic fringe patterns in Columbia resin (CR-39) under explosive loading (camera speed: 193 000 frames per second).

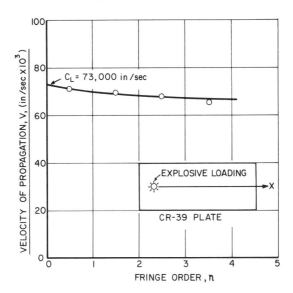

Fig. 11. Wave propagation in Columbia resin (CR-39) plate as a function of fringe order.

193 000 frames per second. The incident P-wave (dilatational) is clearly manifested by the circular arc shape of the fringes. The nearly parallel fringes inclined with respect to the sides of the specimen are produced by the distortional (shear) PS-wave reflected from the free edges of the specimen. The propagation velocity of several fringe orders was computed from this record and plotted versus fringe order in Fig. 11. The propagation velocity of the wave front (zero order fringe) is obtained by extrapolating this curve to the zero order fringe. This measured velocity is the so-called dilatational plate velocity and is related to the bulk dilatational velocity as follows:

$$c_L = \frac{c_1}{1 - \nu^2} \tag{45}$$

Results from tests with a number of model materials are tabulated in Table 1, where the wave propagation velocities and impedance are given. This type of information is useful in selecting model materials of different impedances for studies of wave propagation in layered media.

TABLE 1
Properties of Model Materials for Dynamic Photoelasticity

Material	Wave propagation velocity in plate $c_L (ms^{-1})$	Density $\rho (kg\,m^{-3})$	Impedance $\rho c_L (10^6\,kg\,m^{-2}\,s^{-1})$
Columbia resin (CR-39)	1850	1293	2·392
Polymethyl methacrylate (Plexiglas)	2310	1178	2·721
Polyester (Homalite-100)	2490	1178	2·933
Polyester (Laminac 126-3)	1780	1230	2·189
Polyester (PSM-1)	1650	1146	1·891
Cellulose acetate butyrate	1520	1157	1·759
DER 50/50[a]	1610	1157	1·863
DER 60/40[a]	1040	1146	1·192
DER 70/30[a]	650	1125	0·731
DER 75/25[a]	265	1114	0·295
DER 85/15[a]	147	1110	0·163
Urethane rubber (Hysol)	137	1135	0·155

[a] DER x/y denotes mixture of x-parts of Dow epoxy resin DER 732 and y parts of DER 331.

5.3. Viscoelastic Wave Propagation

In many wave propagation problems related to geophysics a more realistic modelling of the prototype earth medium is made by using a linear viscoelastic model material. Most theoretical studies use Laplace or other transforms in connection with the correspondence principle of viscoelasticity, but assume simple, idealised, mathematically expressible viscoelastic behaviour in performing the necessary transform inversions. Analytical methods exist for treating realistic material properties (Arenz[73]). The experimental approach to the viscoelastic wave propagation problem was described by Daniel using a dynamic photoviscoelastic method.[23,74,75]

Figure 12 shows combined isochromatic and moiré fringe patterns for a viscoelastic plate (PVC) subjected to point impact on the edge. Free-field stresses at a point along the vertical axis of symmetry were obtained independently from moiré and photoelastic data, as follows:

$$\sigma_x(t) = \frac{1}{1-\nu^2} \int_0^t E(t-\tau)\frac{d}{d\tau}[\varepsilon_x(\tau) + \nu\varepsilon_y(\tau)]\,d\tau \tag{46}$$

$$\sigma_y(t) = \frac{1}{1-\nu^2} \int_0^t E(t-\tau)\frac{d}{d\tau}[\varepsilon_y(\tau) + \nu\varepsilon_x(\tau)]\,d\tau \tag{47}$$

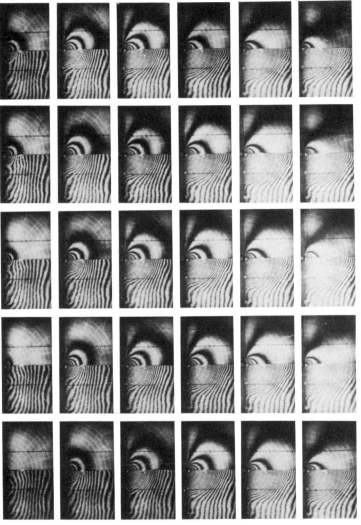

Fig. 12. Series of 30 consecutive photographs showing transient isochromatic and moiré fringe patterns in a plate subjected to impact of a falling weight.[74] (Camera speed: 7400 frames per second; 300 horizontal grid lines per inch).

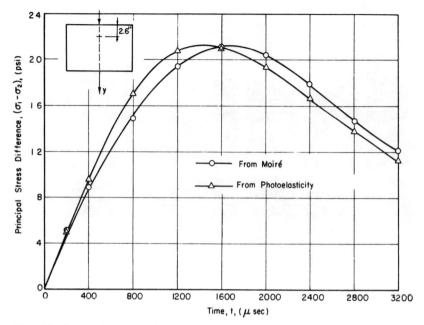

Fig. 13. Comparison of principal stress difference at point 6·6 cm below impacted point of plate obtained independently from moiré and photoelastic data.[74]

from moiré data, and

$$\sigma_x(t) - \sigma_y(t) = \frac{2}{h} \int_0^t f_\sigma(t-\tau) \frac{\mathrm{d}N(\tau)}{\mathrm{d}\tau} \,\mathrm{d}\tau \qquad (48)$$

from photoelastic data, where $E(t) =$ relaxation Young's modulus. Results obtained by the two methods are compared in Fig. 13.

The same procedures were used to study the interaction of a plane viscoelastic wave and a cylindrical cavity. The PVC plate specimen was loaded along one edge with an air shock. Free-field stresses were determined from photoelastic and moiré data. Circumferential stresses around the hole boundary were determined from the photoelastic data obtained from the fringe patterns of Fig. 14 as follows:

$$\sigma_{\theta\theta} = \frac{2}{h} \int_0^t f_\sigma(t-\tau) \frac{\mathrm{d}N(\tau)}{\mathrm{d}\tau} \,\mathrm{d}\tau \qquad (49)$$

These stresses were compared with equivalent 'static' stresses around the hole obtained from the free-field stresses. Results were expressed in the form of stress ratios between the maximum compressive stress on the hole boundary and the maximum compressive stress in the free field (Fig. 15). The most significant result is that the dynamic stress ratio is highest in the beginning, higher than the static ratio by more than 60%.

5.4. Wave Propagation in Layered Media

The problem of wave propagation in layered media, of particular interest in geophysics, has been studied analytically and experimentally. Of special interest is the interaction of stress waves with the interface between two dissimilar media with the resulting generation of reflected, refracted and head waves.

Dynamic photoelasticity was applied successfully by Riley and Dally[76] to the study of wave propagation in a layered model. The model consisted of a Columbia resin (CR-39) and an aluminium plate with an impedance ratio of 6:1. Incident, reflected and head waves in the CR-39 layer were clearly identified and stresses were computed in some cases. Daniel and Marino[62] applied dynamic photoelasticity complemented with moiré to the study of wave propagation in a layered model consisting of two transparent photoelastic layers with an impedance ratio of 2:1. The materials were CR-39 and DER 60/40 mentioned before. The specimens were loaded by detonating a charge of pentaerythritol tetranitrate (PETN) in the low or high impedance layers. The dynamic fringe patterns were photographed with a Cranz–Schardin camera at a rate of approximately 200 000 frames per second.

Figure 16 is one of a sequence of frames obtained for the case of point source loading in the low impedance medium. The fringe pattern shows the incident (P_1) and refracted (P_1P_2) dilatational waves, and the reflected and refracted shear headwaves $(P_1P_2S_1, P_1P_2S_2$ and $P_1S_2S_1)$. Stress separation was achieved for the refracted P_1P_2 wave using only photoelastic data and taking into consideration approximate polar symmetry. Stress separation in the high impedance medium was accomplished by using complementary moiré data. A sequence of frames of moiré fringe patterns is shown in Fig. 17. Stresses computed from photoelastic and moiré data are shown in Fig. 18.

5.5. Wave Propagation in Anisotropic Media

The method of anisotropic photoelasticity (photo-orthotropic elasticity), developed originally for static loading cases, has been extended to

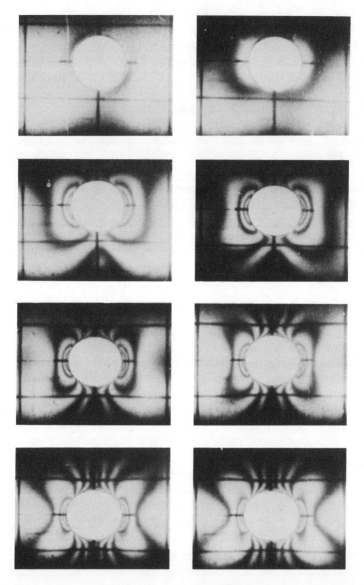

Fig. 14. Series of 16 photographs showing transient isochromatic fringe patterns around a circular hole in a plate subjected to an air shock wave travelling along top edge.[75] (Photographs taken every 255 μs.)

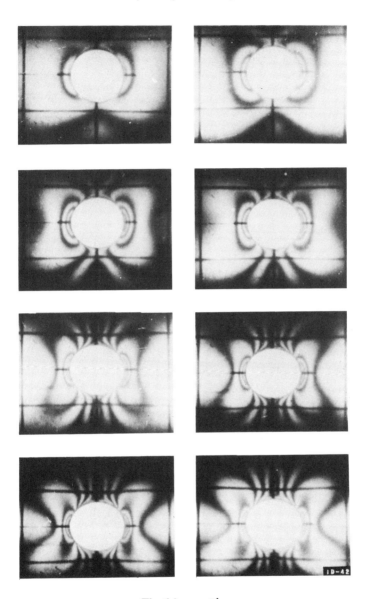

Fig. 14.—*contd.*

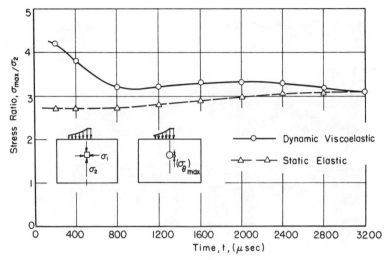

Fig. 15. Ratios of the maximum compressive stress on the hole boundary to the maximum compressive stress in the free-field.[75]

dynamic problems. The method applies transmission photoelastic techniques to stress analysis of birefringent fibrous composite materials. These materials are glass-fibre reinforced plastics with the matrix and the fibres having the same index of refraction. These transparent composites are treated as homogeneous materials with anisotropic elastic and optical properties. The first photoelastic study of wave propagation in transparent composite materials was reported by Dally *et al.*[77] Further applications were described by Rowlands *et al.*[26]

Isochromatic fringe propagation in impacted transparent anisotropic plates is shown in Fig. 19. The plates were made of polyester (Paraplex P444A blended with styrene) reinforced with unidirectional E-glass fibres. The plates were loaded with an explosive charge on one edge, first transverse to the fibre direction and then in the fibre direction. The fringe patterns were recorded with a Beckman and Whitley camera operating at rates of 377 500 and 457 000 frames per second. The five frames shown in Fig. 18 for each case are representative of the 25-frame sequence recorded. The time after impact in microseconds is indicated below each frame. Quasi-dilatational P and quasi-shear S wavefronts are evident, as is the von Schmidt PS wave generated by the P wave moving at grazing incidence along the free

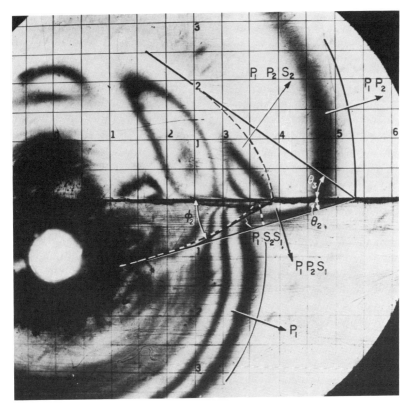

Fig. 16. Wave propagation in layered model illustrating reflected and refracted waves.[63]

boundary. The measured dilatational and shear wave velocities parallel and normal to the fibres are approximately $4570 \ ms^{-1}$ and $2790 \ ms^{-1}$, respectively.

5.6. Dynamic Fracture

One of the most important applications of dynamic photoelasticity is fracture.[27] One of the earliest studies is that by Wells and Post.[15] The major contributors in this area in recent years have been Dally and his associates[78–80] and Kobayashi and his associates.[56,81]

The dynamic fracture behaviour of brittle polymers can be characterised by developing a relation between the instantaneous stress intensity factor and the crack tip velocity. Since the state of stress near

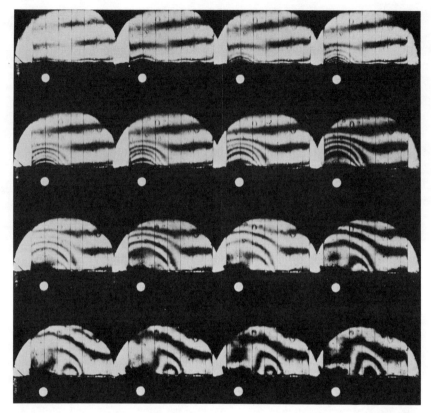

Fig. 17. Moiré fringe patterns for vertical displacements in model 1 (explosive source in low-impedance medium 2·5 cm from interface; camera speed: 190 000 frames per second).[63]

the crack tip is related to the fringe pattern, it is possible to determine the stress intensity factor in terms of fringe pattern parameters. A typical fringe pattern near the tip of a running crack is shown in Fig. 20. Several analysis methods have been proposed. One of the most general relationships between isochromatic fringe parameters and stress intensity factors is the following:[27]

$$(Nf_\sigma/h)^2 = \frac{1}{2\pi r}[(K_I \sin \theta + 2K_{II} \cos \theta)^2 + (K_{II} \sin \theta)^2]$$

$$+ \frac{2\sigma_{0x}}{\sqrt{(2\pi r)}} \sin \frac{\theta}{2}[K_I \sin \theta(1 + 2 \cos \theta)$$

$$+ K_{II}(1 + 2\cos^2 \theta + \cos \theta)] + \sigma_{0x}^2 \qquad (50)$$

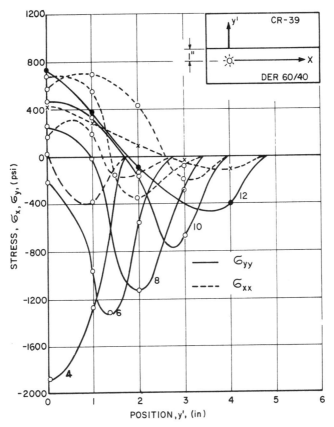

Fig. 18. Principal stresses in high-impedance medium as a function of distance from interface with time as a parameter.[63] (Frame numbers are marked.)

where N = fringe order, f_σ = material fringe value, h = specimen thickness, r = radial distance to point in question, θ = angle from crack axis to point in question, K_I and K_{II} = mode I and mode II stress intensity factors and σ_{0x} = far-field stress along crack direction.

This equation is solved for the unknowns K_I, K_{II} and σ_{0x} by applying it at several data points and using numerical iterative procedures.

Dynamic brittle fracture can be characterised over the entire range from crack arrest to crack branching by the relationship between stress intensity factor K and crack-tip velocity $\dot{a}$. One such relationship is shown in Fig. 21 for a brittle polyester (Homalite-100).[27] The

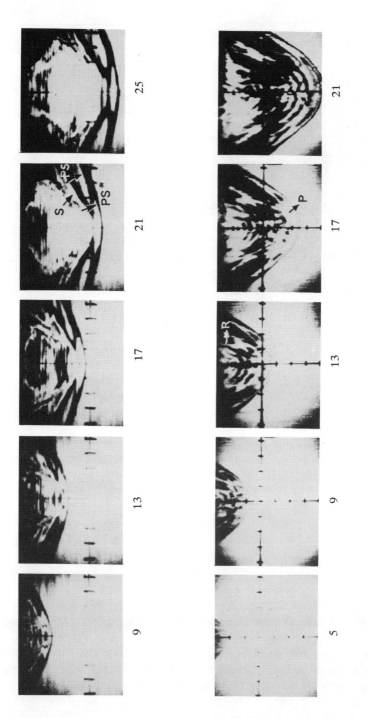

Fig. 19. Isochromatic fringe patterns in glass–polyester composite plate impacted on top edge (Beckman and Whitley camera, electronic flash).[26] (Top) Horizontal fibres, 377 500 fps. (Bottom) Vertical fibres, 457 000 fps.

Fig. 20. Isochromatic fringe pattern around tip of a high-velocity crack in Homalite-100; crack velocity, $\dot{a} = 380$ ms^{-1}. (Courtesy J. W. Dally, and Photomechanics Laboratory, University of Maryland.[80])

minimum and maximum values of K are the arrest toughness and the branching toughness, respectively. Below the minimum value of K no crack propagation takes place. At the highest value of K crack branching occurs. Typical isochromatic fringe patterns around a running crack, before and after crack branching, are shown in Fig. 22.

The latest developments in photoelastic applications to fracture mechanics deal with crack branching and curving. A recently proposed crack branching criterion requires a critical stress intensity factor to trigger crack branching and a crack curving criterion for predicting the crack branching angle.[82,83] These criteria were verified by means of dynamic photoelasticity.

Hybrid techniques employing dynamic photoelasticity and dynamic finite element methods are used effectively in analysing the state of stress near a running crack tip.[84] Dynamic photoelasticity is also used as a means of verifying dynamic finite element codes in wave propagation and fracture mechanics.

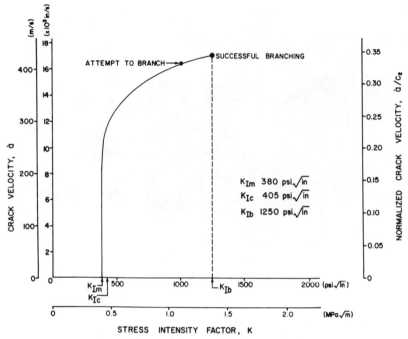

Fig. 21. Crack velocity as a function of the instantaneous stress intensity factor for Homalite-100. (Courtesy J. W. Dally.[27])

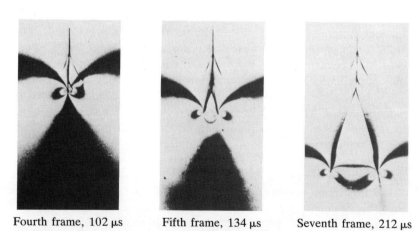

Fourth frame, 102 μs   Fifth frame, 134 μs   Seventh frame, 212 μs

Fig. 22. Typical crack branching dynamic photoelastic fringe patterns in Homalite-100 single-edge-notched specimen. (Courtesy A. S. Kobayashi.[82])

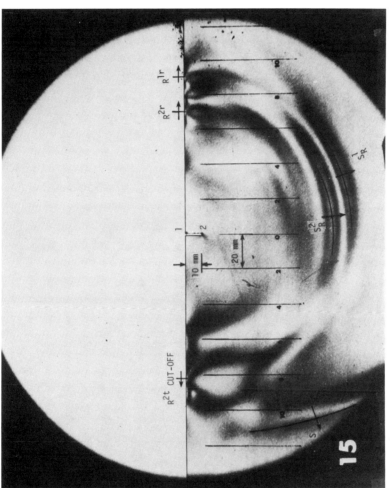

Fig. 23. Reflected, transmitted and mode-converted waves following interaction of Rayleigh wave with slot. (Courtesy C. P. Burger.[85])

5.7. Application to Non-destructive Testing

Dynamic photoelasticity can help visualise the interaction of elastic waves with defects and thus aid in the development and understanding of ultrasonic non-destructive flaw detection. Figure 23 shows an example of the interaction between Rayleigh waves and cracks on the surface of a half-plane.[86] Reflected, transmitted and mode-converted waves are identified. It is shown how the frequency spectrum of the transmitted wave can be used to determine the depth of surface cracks.

5.8. Applications to Water Jet Cutting

The mechanics of fracture by water jet impact have been studied by dynamic photoelasticity.[86,87] A device was used for explosively propel-

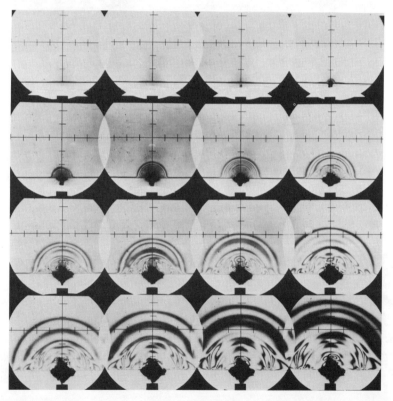

Fig. 24. Isochromatic fringe patterns in 6·4 mm CR-39 specimen impacted with water jet at a velocity of 2500 ms^{-1}. (Camera speed: 180 000 frames per second.[86])

ling water jets at speeds up to 2800 ms^{-1}. The impact of the water jet on transparent photoelastic specimens was studied photoelastically. Model materials used included CR-39, Homalite-100, Plexiglas and glass. Fringe patterns were recorded in the vicinity of the impact with a spark-gap camera at a rate of 180 000 frames per second. A sequence of frames for a CR-39 specimen is shown in Fig. 24. Since the initial pressure build-up is very fast, fringe patterns in the initial stage of impact were also recorded with a pulsed ruby laser at extremely short exposures (15 ns). Figure 25 shows a sequence of such patterns. The fringe patterns show the pressure build-up, wave propagation and crater formation.

The dilatational waves exhibited high attenuation but the shear waves were much more pronounced. Below a threshold level (of 850 ms^{-1} for CR-39) the water jet did not produce noticeable stress waves but resulted in localised jet penetration and cratering, enhanced by the shear wave, followed by quasi-static crater pressurisation resulting in crack opening and extension.

5.9. Application of Photoelastic Coatings

While photoelastic model analysis is an effective means for measuring stresses, its application is limited to problems that can be studied by means of transparent models. Birefringent or photoelastic coatings on the other hand can be readily applied to opaque prototype structures. The method consists of bonding a thin sheet of photoelastic material to the surface of the specimen, such that the bonded interface is reflective. When the specimen is loaded, the surface strains are transmitted to the coating and produce a fringe pattern which is recorded and analysed by means of a reflection polariscope.

Photoelastic coatings have been used to study dynamic fracture in metallic and rock materials.[88-90] Isochromatic fringe patterns illustrating wave propagation in an explosively loaded marble slab are shown in Fig. 26. The marble specimen was coated with a 2 mm thick birefringent coating and loaded with a 110 mg pentaerythritol tetranitrate (PETN) charge. The fringe patterns were recorded with a Beckman and Whitley camera operating at a speed of 500 000 frames per second. The dilatational wave propagation velocity measured from such records was 4060 ms^{-1} for the marble tested. The region of highly dense fringes represents failure of the marble material. A similar marble specimen with a photoelastic coating was loaded with a larger explosive charge (331 mg PETN). A larger time span and a later stage

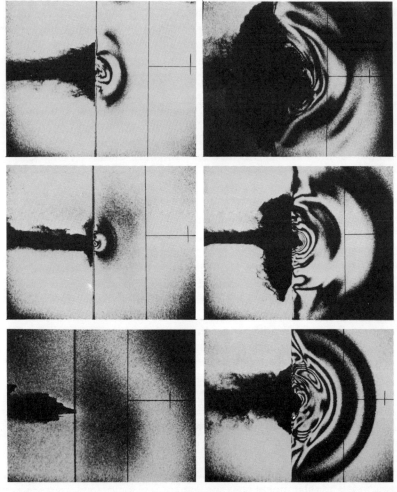

Fig. 25. Sequence of isochromatic fringe patterns for water jet impact on CR-39 specimen at a velocity of $900 \, \mathrm{ms}^{-1}$.[87]

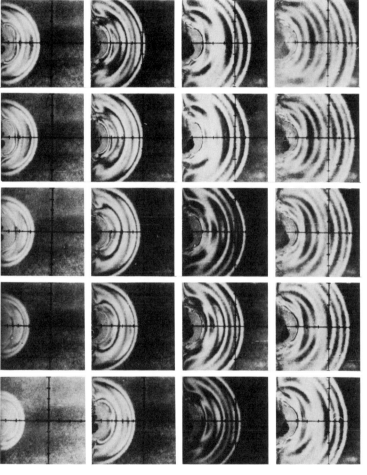

Fig. 26. Isochromatic fringe patterns in photoelastic coating on marble specimen loaded explosively on the edge. (Camera speed: 500 000 frames per second; time range: 18–56 μs after loading.[89])

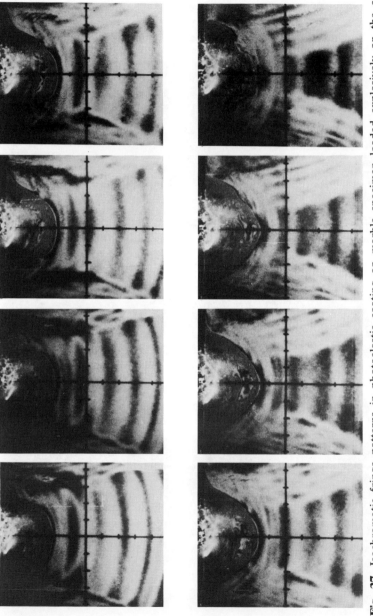

Fig. 27. Isochromatic fringe patterns in photoelastic coating on marble specimen loaded explosively on the edge. (Camera speed: 250 000 frames per second; time range: 70–98 μs.[89])

of the phenomenon were recorded by operating the camera at a rate of 250 000 frames per second. The fringe patterns obtained show clearly the shear waves reflected from the sides of the specimen (Fig. 27). A shear wave velocity of 2540 ms^{-1} was measured. The fractured region is larger than in the previous case.

6. CONCLUSION

An overview was presented of the development and status of dynamic photoelasticity. Considerable progress has been made in all its aspects, development of model materials, loading techniques, recording methods, data analysis and interpretation, and range of applications.

In the area of model materials, photoviscoelastic materials, materials of varying impedances and transparent anisotropic composites have been produced. Loading techniques range from falling weight, to explosive, to water jet loading. The greatest progress noted is in the high speed recording technology. Framing cameras with rates up to 10 million frames per second and 50–100 frame capability are now available. The latest development in spark-gap cameras makes use of individual Q-switched ruby lasers providing intense and ultrashort light pulses, coupled with a microprocessor for synchronisation and control. The most recent development in electronic recording of images is that of the digital imaging camera, based on a digital grey level analyser and picture processing system. This system affords digital recording at very high sensitivity, equivalent to a fringe multiplication of over 500.

The availability of advanced instrumentation and computers allows the acquisition and analysis of data at a higher rate and with more reliability. In the area of application the emphasis has been on fracture and damage mechanics. The interaction between theory and dynamic photoelasticity is becoming stronger as new hybrid numerical-experimental approaches are introduced.

REFERENCES

1. Tuzi, Z., Photographic and kinematographic study of photoelasticity, *Scientific Papers, Inst. of Phys. and Chem. Research*, **8** (149) (1928), 247–67.
2. Tuzi, Z. and Nisida, M., Photoelastic study of stresses due to impact, *Philosophical Magazine, Ser. 7*, **21** (1936), 448–73.

3. Findley, W. N., The foundations of photoelastic stress analysis applied to dynamic stresses, *Proc. 9th Semi-annual Photoelasticity Conf.*, 1939, pp. 1–11.

4. Senior, D. A. and Wells, A. A., A photoelastic study of stress waves, *Philosophical Magazine, Ser.* 7, **37** (1946), 463–9.

5. Foeppl, L. Slow motion pictures of impact tests by means of photoelasticity, *J. Appl. Mech.*, **71** (1949), 173–7.

6. Christie, D. G., An investigation of cracks and stress waves in glass and plastics by high-speed photography, *J. Soc. Glass Tech.*, **36** (1952), 74–89.

7. Goldsmith, W., Dynamic photoelasticity, In: *Experimental Techniques in Shock and Vibration*, ed. W. J. Worley, ASME, New York, 1962, pp. 25–54.

8. Dally, J. W., An introduction to dynamic photoelasticity, *Exp. Mech.*, **20** (12) (1980), 409–16.

9. Perkins, H. C., Movies of stress waves in photoelastic rubber, *J. Appl. Mech.*, **20** (1953), 140–1.

10. Dally, J. W., Riley, W. F. and Durelli, A. J., A photoelastic approach to transient stress problems employing low-modulus materials, *J. Appl. Mech.*, **26** (1959), 613–20.

11. Feder, J. C., Gibbons, R. A., Gilbert, J. T. and Offenbacher, E. L., The study of the propagation of stress waves by photoelasticity, *Proc. SESA*, **XIV** (1) (1956), 109–22.

12. Flynn, P. D., Feder, J. C., Gilbert, J. T. and Roll, A. A., Some new techniques for dynamic photoelasticity, *Exp. Mech.*, **2** (1962), 159–60.

13. Frocht, M. M., Flynn, P. D. and Landsberg, D., Dynamic photoelasticity by means of streak photography, *Proc. SESA*, **XIV** (2) (1957), 81–90.

14. Christie, D. G., A multiple spark camera for dynamic stress analysis, *J. Photogr. Science*, **3** (1955), 153–9.

15. Wells, A. A. and Post, D., The dynamic stress distribution surrounding a running crack—a photoelastic analysis, *Proc. SESA*, **XVI** (1) (1958), 69–96.

16. Riley, W. F. and Dally, J. W., Recording dynamic fringe patterns with a Cranz–Schardin camera, *Exp. Mech.*, **9** (8) (1969), 27N–33N.

17. Rowlands, R. E., Taylor, C. E. and Daniel, I. M., A multiple-pulse ruby laser system for dynamic photomechanics. Applications to transmitted and scattered-light photoelasticity, *Exp. Mech.*, **9** (9) (1969), 385–93.

18. Dally, J. W. and Sanford, R. J., Multiple ruby laser system for high speed photography, *Optical Engineering*, **21** (4) (1982), 704–8.

19. Dally, J. W., Data analysis in dynamic photoelasticity, *Exp. Mech.*, **7** (8) (1967), 332–8.

20. Flynn, P. D., Dual-beam polariscope and framing camera for dynamic photoelasticity, *Exp. Mech.*, **13** (4) (1973), 178–84.

21. Post, D., A new photoelastic interferometer suitable for static and dynamic measurements, *Proc. SESA*, **XII** (1) (1954), 191–201.

22. Bohler, P. and Schumann, W., On the complete determination of dynamic states of stresses, *Exp. Mech.*, **8** (3) (1968), 115–21.

23. Daniel, I. M., Experimental methods for dynamic stress analysis in viscoelastic materials, *J. Appl. Mech., Series E*, **32** (1965), 598–606.

24. Cole, C. A., Quinlan, J. F. and Zandman, F., The use of high-speed photography and photoelastic coatings for the determination of dynamic strains, In: *Proc. Fifth Intern. Congress on High-Speed Photography*, ed. J. S. Courtney-Pratt, SMPTE, New York, 1962, pp. 250–61.

25. Dally, J. W. and Riley, W. F., Initial studies in three-dimensional dynamic photoelasticity, *J. Appl. Mech.*, **34** (1967), 405–10.

26. Rowlands, R. E., Daniel, I. M. and Prabhakaran, R., Wave motion in anisotropic media by dynamic photomechanics, *Exp. Mech.*, **14** (11) (1974), 433–9.

27. Dally, J. W., Dynamic photoelastic studies of fracture, *Exp. Mech.*, **19** (10) (1979), 349–61.

28. Mindlin, R. D., A mathematical theory of photoviscoelasticity, *J. Appl. Phys.*, **20** (1949), 206–16.

29. Read, W. T., Jr., Stress analysis for compressible viscoelastic materials, *J. Appl. Phys.*, **21** (1950), 671–4.

30. Dill, E. H., A theory for photothermoviscoelasticity, *Intern. Congress for Rheology*, Brown University, Providence, Rhode Island, 1963.

31. Williams, M. L. and Arenz, R. J., The engineering analysis of linear viscoelastic materials, *Exp. Mech.*, **4** (1964), 249–62.

32. Khesin, G. L., Kostin, I. K., Roshdestvenski, K. N. and Shpyakin, V. N., Techniques and results of calibration studies of birefringent epoxy polymers under dynamic loading, *Model Studies of Dynamic, Thermal, and Static Photoelastic Methods*, Moscow, 1970. (In Russian.)

33. Coleman, B. D. and Dill, E. H., Photoviscoelasticity: theory and practice, In: *The Photoelastic Effect and Its Applications*, ed. J. Kestens (Intern. Union of Theoretical and Applied Mechanics), Springer-Verlag, Berlin, 1975, pp. 455–505.

34. Doyle, J. F., An interpretation of photoviscoelastic/plastic data, *Exp. Mech.*, **20** (2) (1980), 65–7.

35. Clark, A. B. J., Static and dynamic calibration of a photoelastic model material, CR-39, *Proc. SESA*, **14** (1) (1956), 195–204.

36. Clark, A. B. J. and Sanford, R. J., A comparison of static and dynamic properties of photoelastic materials, *Proc. SESA*, **20** (1) (1963), 148–51.

37. Frocht, M. M., Studies in dynamic photoelasticity with special emphasis on the stress-optic law, *Int. Symposium on Stress Wave Propagation in Materials*, Interscience Publishers, New York, 1960, pp. 91–118.

38. Daniel, I. M., Dynamic properties of a photoviscoelastic material, *Exp. Mech.*, **6** (5) (1966), 225–34.

39. Brown, G. W. and Selway, D. R., Frequency response of a photoviscoelastic material, *Exp. Mech.*, **4** (3) (1964), 57–63.

40. Chase, K. W. and Goldsmith, W., Mechanical and optical characterization of an anelastic polymer at large strain rates and large strains, *Exp. Mech.*, **14** (1) (1974), 10–18.

41. Peeters, R. L. and Parmerter, R. R., Optical calibration of photoviscoelastic materials on a microsecond time scale, *Exp. Mech.*, **14** (11) (1974), 445–51.

42. Ninomiya, K. and Ferry, J. D., Some approximate equations useful in the phenomenological treatment of linear viscoelastic data, *J. Colloid. Science*, **14** (1959), 36–48.

43. Sackman, J. L. and Kaya, I., On the determination of very early-time viscoelastic properties, *J. Mech. Phys. Solids*, **16** (1968), 212–32.
44. Clark, J. A. and Durelli, A. J., An introduction to dynamic photoelasticity: Discussion of paper by J. W. Dally (*Exp. Mech.*, **20** (1980), 409–16), *Exp. Mech.*, **23** (1) (1983), 42–8.
45. Rowlands, R. E., Taylor, C. E. and Daniel, I. M., Ultrahigh-speed framing photography employing a multiply-pulsed ruby laser and a 'smear-type' camera: application to dynamic photoelasticity, In: *High-Speed Photography*, ed. N. R. Nilsson and L. Hogberg, John Wiley and Sons, New York, 1968, pp. 275–80.
46. Clark, J. A. and Durelli, A. J., Optical stress analysis of flexural waves in a bar, *J. Appl. Mech.*, **37** (1970), 331–8.
47. Durelli, A. J. and Riley, W. F., Stress distribution on the boundary of a circular hole in a large plate during passage of a stress pulse of long duration, *J. Appl. Mech.*, **83** (1961), 245–51.
48. Daniel, I. M. and Riley, W. F., Stress distribution on the boundary of a circular hole in a large plate due to an air shock wave traveling along an edge of the plate, *J. Appl. Mech.*, **31** (3) (1964), 402–8.
49. Flynn, P. D., Photoelastic studies of dynamic stresses in high modulus materials, *99th Tech. Conf. SMPTE*, 1966.
50. Lieber, A. J. and Sutphin, H. D., Nanosecond high resolution framing camera, *Review of Sci. Instruments*, **42** (11) (1971), 1663–7.
51. Cranz, C. and Schardin, H., Kinematographic auf ruhendem Film und mit extrem hoher Bildfrequenz, *Zeit. f. Phys.*, **56** (1929), 147.
52. Conway, J., An improved Cranz–Schardin high-speed camera for two-dimensional photomechanics, *Review of Sci. Instruments*, **43** (8) (1972), 1172–4.
53. Hendly, D. R., Turner, J. L. and Taylor, C. E., A hybrid system for dynamic photoelasticity, *Exp. Mech.*, **15** (8) (1975), 289–94.
54. Becker, H., Simplified equipment for photoelastic studies of propagating stress waves, *Proc. SESA*, **18** (2) (1960), 214–16.
55. Tschinke, M., Studio fotoelastico sui fenomeni transitori all'attaco delle palette di turbomachine, *Tecnica Italiana*, **XXXI** (11) (1966).
56. Bradley, W. B. and Kobayashi, A. S., An investigation of propagating cracks by dynamic photoelasticity, *Exp. Mech.*, **10** (3) (1970), 106–13.
57. Schwieger, H., Photoelastic study of the impact of thin glass rods, In: *Photographie et Cinematographie Ultrarapides*, eds Nasline and Vivie, Dunod, Paris, 1956, pp. 345–51.
58. Cunningham, D. M., Brown, G. W. and Griffith, J. C., Photoelastometric recording of stress waves, *Exp. Mech.*, **10** (3) (1970), 114–19.
59. Voloshin, A. S. and Burger, C. P., Half-fringe photoelasticity: a new approach to whole-field stress analysis, *Exp. Mech.*, **23** (3) (1983), 304–13.
60. Durelli, A. J., Dally, J. W. and Riley, W. F., Developments in the application of the grid method to dynamic problems, *J. Appl. Mech.*, **26** (1959), 629–34.
61. Riley, W. F. and Durelli, A. J., Application of moiré methods to the determination of transient stress distributions, *J. Appl. Mech.*, **29** (1) (1962), 23–9.

62. Daniel, I. M. and Marino, R. L., Wave propagation in layered model due to point-source loading in low-impedance medium, *Exp. Mech.*, **11** (5) (1971), 210–16.

63. Favre, H., Sur une nouvelle méthode optique de determination des tensions interieures, *Rev. d'Optique*, **8** (1929), 193–213, 241–61, 289–307.

64. Post, D., Photoelastic evaluation of individual principal stresses by large field absolute retardation measurements, *Proc. SESA*, **XIII** (2) (1956), 119–32.

65. Nisida, M. and Saito, H., A new interferometric method of two-dimensional stress analysis, *Exp. Mech.*, **4** (12) (1964), 366–76.

66. Fourney, M. E., Applications of holography to photoelasticity, *Exp. Mech.*, **8** (1) (1968), 33–8.

67. Hovanesian, J. D., Brcic, V. and Powell, R. L., A new experimental stress-optic method: stress-holo-interferometry, *Exp. Mech.*, **8** (8) (1968), 362–8.

68. Clark, J. A. and Durelli, A. J., A simple holographic interferometer for static and dynamic photomechanics, *Exp. Mech.*, **10** (12) (1970), 497–505.

69. Sanford, R. J. and Durelli, A. J., Interpretation of fringes in stress-holo-interferometry, *Exp. Mech.*, **11** (4) (1971), 161–6.

70. Holloway, D. C., Ranson, W. F. and Taylor, C. E., A neoteric interferometer for use in holographic photoelasticity, *Exp. Mech.*, **12** (10) (1972), 461–5.

71. Dudderar, T. D. and Doerries, E. M., A holographic interferometer for dynamic photoelasticity, *ASME Paper 74-DE-14*, 1974.

72. Lallemand, J. P. and Lagarde, A., Separation of isochromatics and isopachics using a Faraday rotator in dynamic-holographic photoelasticity, *Exp. Mech.*, **22** (5) (1982), 174–9.

73. Arenz, R. J., Two dimensional wave propagation in realistic viscoelastic materials, *J. Appl. Mech.*, **32** (1965), 303–14.

74. Daniel, I. M., Stresses around a circular hole in a viscoelastic plate subjected to point impact on one edge, In: *Developments in Mechanics*, Vol. 3, Part I, ed. T. C. Huang and M. W. Johnson, Jr., John Wiley and Sons, New York, 1966, pp. 491–547.

75. Daniel, I. M., Viscoelastic wave interaction with cylindrical cavity, *J. Exp. Mech. Division, ASCE*, **92** (EM6) (Proc. Paper 4999) (1966), 25–42.

76. Riley, W. F. and Dally, J. W., A photoelastic analysis of stress wave propagation in a layered model, *Geophysics*, **31** (1966), 881–9.

77. Dally, J. W., Link, J. A. and Prabhakaran, R., A photoelastic study of stress waves in fiber reinforced composites, *Proc. 12th Midwestern Mechanics Conf.*, 1971, 937–49.

78. Kobayashi, T. and Dally, J. W., The relation between crack velocity and the stress intensity factor in birefringent polymers, *ASTM STP 627*, 1977, 257–73.

79. Sanford, R. J. and Dally, J. W., A general method for determining mixed-mode stress intensity factors from isochromatic fringe patterns, *J. Eng. Fract. Mech.*, **11** (1979), 621–33.

80. Dally, J. W., Developments in photoelastic analysis of dynamic fracture,

Proc. IUTAM Symposium on Optical Methods in Mechanics of Solids, Sijthoff and Noordhoff, The Hague, 1981, pp. 359–94.
81. Kobayashi, A. S. and Mall, S., Dynamic fracture toughness of Homalite-100, *Exp. Mech.*, **18** (1) (1978), 11–18.
82. Ramulu, M., Kobayashi, A. S. and Kang, B. S.-J., Dynamic crack branching—A photoelastic evaluation, *Fracture Mechanics—Fifteenth Symposium, ASTM STP 833*, 1984, 103–48.
83. Ramulu, M. and Kobayashi, A. S., Dynamic crack curving—A photoelastic evaluation, *Exp. Mech.*, **23** (1) (1983), 1–9.
84. Kobayashi, A. S., Seo, K., Jou, J. Y. and Urabe, Y., A dynamic analysis of modified compact-tension specimens using Homalite-100 and polycarbonate plates, *Exp. Mech.*, **20** (3) (1980), 73–9.
85. Burger, C. P., Testa, A. and Singh, A., Dynamic photoelasticity as an aid in developing new ultrasonic test methods, *Exp. Mech.*, **22** (4) (1982), 147–54.
86. Daniel, I. M., Rowlands, R. E. and Labus, T. J., Photoelastic study of water jet impact, *Second International Symposium on Jet Cutting Technology*, BHRA, Cambridge, 1974, paper A1.
87. Daniel, I. M., Experimental studies of water jet impact on rock and rocklike materials, *Third International Symposium on Jet Cutting Technology*, BHRA, Chicago, 1976, paper B3.
88. Van Elst, H. C., *Dynamic Crack Propagation*, ed. G. C. Sih, Noordhoff International Publishing, Leyden, 1972, pp. 283–332.
89. Daniel, I. M. and Rowlands, R. E., On wave and fracture propagation in rock media, *Exp. Mech.*, **15** (12) (1975), 449–57.
90. Kobayashi, T. and Dally, J. W., Dynamic photoelastic determination of à–K relation for 4340 alloy steel, *ASTM STP 711*, 1980, 189–210.

2

PHOTOPLASTICITY AND ITS ROLE IN THE METHODS OF PHOTOMECHANICS

Jan Javornický

Institute of Theoretical and Applied Mechanics, Czechoslovak Academy of Sciences, Prague, Czechoslovakia

ABSTRACT

Inelastic deformation is a consequence of structural changes in materials. These changes do not involve the mechanical theories of plasticity or viscoelasticity which are merely descriptive. Using the photomechanical effect on model materials, however, an understanding of the relation between its initial cause and the subsequent manifestation as a deformation is necessary. The chapter begins with a study of theoretical deformation and its physical meaning; then the physicochemical background of birefringence (and its dispersion) in plastics is dealt with. The theory of photoplasticity is based on the double nature of birefringence which is expressed by the phenomenological eqn (5) and formulated in its principal relation in eqn (11). From the above points of view, the model solution of problems of viscoelastic deformation is then presented together with other approaches. Experimental procedures of investigation and applications of the method to different problems are studied next. Finally, special branches of photoplasticity, i.e. measurements using birefringent coatings and measurements on transparent polycrystalline materials, complete the treatise.

1. SURVEY OF PHOTOPLASTIC METHODS

Among the methods of photoplasticity three different procedures can be distinguished. The first is photoplasticity in the narrow sense of the word, meaning the procedure using transparent plastic models and investigating them as in traditional photoelasticity. It is based on the use of birefringence at stresses beyond the elastic limit, and its theory and techniques are the major content of this chapter.

The second method is the use of polycrystalline model materials. Its techniques are in principle the same as in the previous method. Its merit lies in the similarity of structure of the model material to that of metals, a disadvantage being its very low yield point. This method is treated in the last part of the chapter.

The third method measures plastic strains on the surfaces of structural elements and bodies by means of birefringent coatings. Its techniques do not differ greatly from the usual applications of stress-optical coatings and therefore only a brief survey is given.

The principles and techniques of photoplasticity are stressed in the present chapter and its development traced from recent works. For a more detailed description the reader is referred to ref. 1.

2. APPROACHES TO THE THEORIES OF PLASTICITY AND VISCOELASTICITY

The deformation process is a complex phenomenon. The structural changes manifest themselves by various changes in the behaviour of materials stressed gradually up to progressively higher levels. In order to better understand this behaviour, it is necessary to analyse the process.

In mechanics, deformation is understood to be, in physical terms, the relative displacement of parts or particles from their positions in a solid such that the continuity of the solid is not impaired. The displacement of the particles can proceed in such a way that the time of the process, and the time preceding the analysis, will not be decisive for the analysis of the final state. Even in some partial descriptions of processes in which time does play a certain role, the time effect may be neglected. These scleronomous processes of deformation are characteristic of perfect solids. Real materials lack such perfection, however, and time is a factor to be considered in the period of action of external forces, or in the history of loading which the material being studied undergoes. Processes of this type are generally denoted as rheonomous. In practice this means that structural materials exhibit a more or less pronounced property which is characteristic of liquids, i.e. liquidity. This, however, is a transient state of matter which can be achieved by external conditions, e.g. by temperature or by a field of force. In liquids, the simplest case is for Newtonian flow. Under the action of an external force, strain increases continuously and irreversibly with time. Every new position of a material particle constitutes a new equilibrium, and no external force is required to retain the new shape. As soon as one begins to talk of 'flow', the event becomes, inevitably, a rheonomous one. Such a system, however, is not able to accumulate energy. The work expended during deformation is irreversible and the energy supplied is fully converted to heat, resulting in a dissipative system.

Two simple kinds of time-dependent deformation are the basic types of flow. In viscous flow, the dissipated mechanical energy depends upon the strain rate. In plastic flow, or yielding, which is also associated with dissipation since it leads to permanent strain, the energy of dissipation is independent of the rate of strain. In perfectly plastic materials, flow can be maintained by the action of a constant force, whereas to maintain plastic flow in a strain-hardening material an increase in the acting force is necessary.

The development of strain and the history of stress or energy variations prior to the final state cannot be inferred from the latter. Since such time-dependent deformation cannot be described in a physically consistent way, a classification which takes into account the share of the individual contributions must be introduced. Every real material exhibits both an elastic and a time-dependent component of the final strain; the proportion of each is influenced by temperature and the intensity of the stress.

Deformation in which the properties of the perfect solid and of the liquid phase combine is denoted as viscoelastic. In its most simple form it is characterised by a linear relationship between the time function of stress and the linear relationship between the time function of stress and that of strain, enabling the preservation of Boltzmann's principle of superposition. The behaviour of a large number of structural materials fits this description quite well. Besides, a number of them are, in practice, considered to be elastic, and it is only after a closer inspection that their viscoelastic character is revealed.

The need to classify the various types of deformation thus leads to their distinction according to their dependence upon time. Most of the time deformations must be ascribed to viscoelasticity, which possesses a certain component of the character of viscous flow. Theoretically this means that part of the strain is dependent on the rate of shear strain reached, or that the energy of dissipation is dependent on the square of that rate. The proportion of this component then determines the degree of linear or non-linear behaviour.

Thus, one can differentiate between anelasticity, or retarded elasticity, elastic–plastic deformation, viscoelastic deformation, combined elastic and viscoplastic deformation, combined elastic–plastic and anelastic deformation, etc. (see Fig. 1).

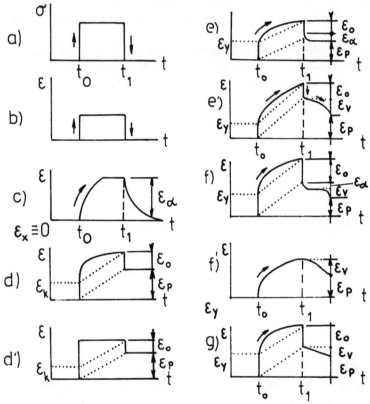

Fig. 1. (a) Loading diagram for all cases; (b) elastic deformation; (c) anelastic deformation; (d) elastic-plastic deformation; (d′) elastic-plastic deformation and strain hardening; (e) combined elastic-plastic and anelastic deformation; (e′) viscoelastic deformation; (f) combined viscoelastic and anelastic deformation; (f′) viscoplastic deformation; (g) combined viscoelastic and elastic deformation. ε_o = elastic component, ε_p = plastic (permanent) component, ε_v = viscous component, ε_α = anelastic component, ε_y = yield strain (limit of elasticity).

From this classification the principal difference between viscoelastic and elastic-plastic deformation can be seen. While the plastic material only suffers permanent deformation when a certain specific value of stress—the yield point—has been exceeded, the viscoelastic material suffers permanent deformation from the very onset of stressing. Thus, these types of deformation do not vary according to the magnitude of the strain achieved, nor to the rate of the deformation, but solely to the magnitude of the stress that provokes permanent stress (strain).

Though in a viscoelastic material there remains a certain residue of permanent strain, the deformation cannot be considered plastic. The physical nature of these two types of deformation is quite different, as explained below. This is why it is quite adequate to only speak about plastic deformation with reference to metals.

Besides static and dynamic time loading, there are two types of time loading which are of theoretical and practical importance. These are loading by constant stress, which brings about a process of strain denoted as creep, and loading whilst maintaining a constant magnitude of strain, which produces a change of stress with time denoted as relaxation. These two types are of importance both for supplying the parameters of the material necessary for a mathematical description of the process of time deformation, and for the study of the mechanism of deformation.

It follows from what has been said that one cannot determine the type of deformation process or its individual components purely from a consideration of the time dependence of strain on creep loading; the nature of the phenomenon lies in changes going on in the structure of the material, and this requires investigation by methods associated with the physics of solids.

The process of creep deformation, despite its non-complex character, throws light upon certain features of mechanical deformation, since it enables decisive changes in the structure to develop. In the case of abrupt and varying loading the time factor is short, and information concerning the nature of the deformation is supplied by other means, e.g. by damping or from the phase shift with time.

The time/strain diagram also yields information about the properties of materials that are, regardless of the possibility of a different structural origin in different materials, in general denoted as recovery. It covers the period of retreat of anelastic deformation after unloading from the state that was reached by elastic–plastic or viscoelastic deformation.

The basic feature of plastic deformation is considered to be the irreversible displacement of structural particles of the material medium. Plastic deformation of the material is a process which results in a change in the shape and position of structural particles relative to their original positions. This process is irreversible, the medium preserving its continuity and retaining its compatibility, with the exception of changes in microstructure, e.g. dislocations.

The definition ties up well with the nature of the deformation process in plastics, which develop a straightening of their molecular chains; it also covers the deformation process in crystalline materials, which deform through distortions and kinking of shear planes, and it also applies to the deformation of inhomogeneous materials in which the individual components participate in the attainment of the resulting state. The irreversibility of the process implies plastic deformation 'from below', i.e. it determines the boundary between the elastic and the plastic state denoted as the yield point, and 'from above', if the material makes it necessary to determine the boundary of the region leading to rupture. This region is characterised by the fact that, besides the relative displacements of the particles, forces that have so far been idle (e.g. valency bonds) now begin contributing to the resistance of the material, and a process initiates which results in the separation of the particles. This manifests itself in metals, for example, by pronounced pinching, cracks, increases in strain rate, etc.

Even though permanent deformation can occur with viscoelastic deformation, it may not be plastic, especially as far as macromolecular materials are concerned. Here the molecules can, for example, extend and remain in the deformed shape, without changing their position, purely because of the constraint elicited by the adjoining molecules. A more frequent case would be a combination of both types, particularly with large strains, but to distinguish between them in these materials would be a matter of chemical physics, and would not be practically expedient.

It has been seen that the deformation of solids is linked with changes in their structure, and photoviscoelasticity and photoplasticity are experimental methods, in which the mechanical phenomena in bodies are simulated by transparent models made of solid macromolecular materials. The mechanical behaviour of these models can be interpreted in terms of optical effects and this is a source of certain limitations. If these methods are applied in an investigation, the aim of this investigation must be formulated explicitly: whether the investigation will be concerned with changes in the structure or only with strains or stresses in the deformation process. That is also why the current viscoelastic or plastic analyses consider strains only in dependence upon loads, although it is the structural, changes that are the main interest of the designer nowadays. Besides, it is essential to know the nature and causes of birefringence in order to be able to interpret correctly its magnitude in terms of strain and stress. To neglect this fact is to induce considerable errors into measurement and evaluation. A knowledge of the properties of the model materials is a crucial precondition for an adequate application of these methods.

The basic groups of polymers are as follows: the linear polymers, which exhibit a fibrous structure of chained molecules; the crosslinked polymers, which consist of a network of simple, chained, or branched molecules; and those polymers which display a well-ordered near-crystalline structure and exhibit a certain degree of amorphousness. Different types of structure suffer deformation in different ways. From a mechanics point of view, a schematic model of their deformation which comprises of two parts is acceptable: an elastic part (solid frame), and a viscous part (liquid filler). This is why the macromolecular materials are taken as the most adequate models of viscoelastic deformation and why they have stimulated development of the theory of deformation.

Plastics (elastomers and plastomers) display permanent deformations (which can be called plastic in the literal sense of the word) and time-dependent deformations which are influenced by the viscous phase. Time-dependent deformations can, after a long period of straining, exceed the first type so that the latter becomes negligible.

3. MODEL MATERIALS AND TIME-DEPENDENT DEFORMATION

3.1. Deformation Behaviour of Plastics

Model materials are solid at room temperature and are capable of temporary birefringence. They are distinguished by a certain optical transparency at specific wavelengths.

Macromolecular materials, as the term indicates, are formed by molecules of a size which, in general, surpasses that of the molecules of inorganic materials. The former are produced by the combination of simpler units by polymerisation, polycondensation and polyaddition.

From the mechanical and optical viewpoint, it is important to note that the end-products of such chemical processes are materials made up of so-called chain molecules, the length and the branching of which can even differ for the same material. Thus the composition and structure of polymers will not be uniform in particular cases; the large number of the various bonds and types of branching results in a considerable variation of structure.

Polymeric molecules, as chemical compounds, can have either ionic bonds (heteropolar), the dissociation of which yields ions, or covalent bonds (homopolar), which yield electroneutral atoms. The molecules of the greater part of polymers have homopolar bonds, which accounts for the dielectric nature of polymers characterised by the dipole moment and polarisability; the latter can be described by an ellipsoid, which is of the tensor type and is directly accountable for the birefringent effect under mechanical loading. The molecules are mainly bound together by chemical bonds and van der Waals forces, which arise between chemically unbound atoms and are produced by the field of rotating free electrons.

Since a precise definition of the polymeric structure is unattainable, such a structure is denoted as amorphous, although it may—in different types—display a certain level of arrangement. The simplest is the linear polymer, which is an unarranged structure having a random dispersion of chains; a higher form is the globular structure in which the molecular chains form wicks, which are ball-like, for example in Teflon. In some polymers, systems of molecules are created whose branches are interlocked and bonded, forming a kind of net. These are the crosslinked polymers. The best-arranged structure is one which is analogous to a crystal structure; the simpler molecular components are spaced out at regular intervals roughly at points which, in the crystal lattice, are reserved for atoms. This structure can thus be considered as being of the crystal type, though a number of characteristics pertaining to an atomic structure of this type are absent.

Examination by X-ray and by polarised light have shown that the group of amorphous polymers can include phenolformaldehyde, polymethylmethacrylate, PVC, polyacetates and celluloid, the group of slightly crystalline polymers, e.g. polystyrene, and the crystalline group, e.g. polyamide, polyethylene, polytetrafluoroethylene (Teflon), etc. Although the different polymers may be classified according to different systems of structure, the basic features of their deformation mechanism can, and very often do, display similarity.

The structure of an amorphous, linear polymer is formed by a skeleton of molecular nuclei loosely interlinked by rather long chains of the same length. The molecular nuclei are affected, on the one hand, by van der Waals mass forces, dipole interactions, hydrogen bonds, labile and stable covalent bonds, etc., and on the other hand, by forces that arise from the tendency of chains to return after deformation to a state of maximum entropy by means of Brownian, or thermal, motion. The presence of bonds of different strength serves as a pointer to the viscoelastic properties of polymers and accounts for the existence of spectra for the relaxation times. Crystalline polymers are, of course, more complex than linear polymers; their behaviour is also different. The transition between the glass and the rubber range is not so distinct, and the complexity of bonds prevents irreversible molecular flow at higher temperatures. Elastomers form a special group. They are linear, yet even, three-dimensionally crosslinked polymers in which the elastic forces acting in the structure are dependent upon a change in the entropy of the whole system, not upon the internal energy; thus the stress and birefringence are linearly related throughout the whole deformation range, and the internal energy is independent of the extent of elongation.

The mechanical properties of polymers are not directly linked to the type of macromolecule; these and other properties are derived from the character of the frequencies and the nature of their vibrations, the angular velocities of their rotation and the internal rotation of the molecules. The forces conditioning these movements are of an electromagnetic nature. The thermodynamic properties of molecules, on the other hand, are linked with the geometrical and purely mechanical relationships in the system. The thermodynamic and kinetic properties of linear polymers can thus be basically derived from the internal rotation of individual links in the molecular chain. A displacement in the rotation of the particles results in a change in the configuration of the macromolecules which can be considered to be deformation. This change consists essentially of rotational isomerism, which in linear polymers almost changes into the orientation of the individual molecules. The influence of the size of molecules, and of the crosslinkage, is stronger at lower frequencies, whereas at high frequencies the type of the atomic bond gains influence. Physico-chemical knowledge concerning macromolecules has not, however, so far reached a level to explain all the complex manifestations.

In polymers, loading causes deformations which, due to internal stress, lead to a new (elastic) equilibrium, and deformations which, owing to stationary reactions, keep growing (flow). In real polymers these types form an inseparable bond. This two-fold aspect of deformation in a rigid polymer, considering linear behaviour only, is a consequence of mechanical reactions arising from two sources: reactions that are energy-elastic, corresponding to Hooke's law, and those that are entropy-elastic ('reversible'), i.e. based on the mechanism of entropy, as in an ideal gas.

On the basis of the analysis of the kinetic mechanism in polymers, three types of deformation can be well distinguished. The three types are easily recognised. If a polymer is mechanically stressed, the chains forming a net undergo a change from a configuration with random distribution to one exhibiting lower entropy. *Elastic deformations* occur at low stress, which causes the chains to disengage by overcoming the van der Waals forces and to open the valency angles, though the valency bonds suffer only mild stress. *Anelastic deformation* develops due to changes in configuration during the process. This occurs very slowly since there is extensive slipping of the molecules over one another. Such changes are, of course, reversible. Finally, *plastic deformation* (or irreversible viscous flow) is governed by whole polymeric chains slipping over each other in the molecular structure.[11,31]

This process covers both mechanical and chemical reactions; one bond is shattered, another, of the same type, is established, etc. The total reaction of the structure is thus a process in which chains are breaking as well as new ones forming. This process is also accompanied by a drop in stress since, when a chain is destroyed, stress relaxes until new bonds are formed. The relationship between straining and changes in structure may be influenced by supermolecular structure.

Under rapid (shock) loading at room temperature, structural changes cannot fully develop; only the spacing of the nuclei of the polymer molecules changes in the same way as in reversible deformation. Under dynamic loading, however, through vibration and under the influence of heat, changes in structure and superstructure do occur even in crystalline polymers, such as reorientation of crystals and flow in the amorphous component or of the crystals themselves and similar events. Under slow or long-term loading the distances between the majority of chain-ends change, the chains elongate in the direction of the acting force and plastic flow ensues. Recent investigations have reconfirmed this explanation.[26]

This behaviour has revealed test graphs as we know them in prac-

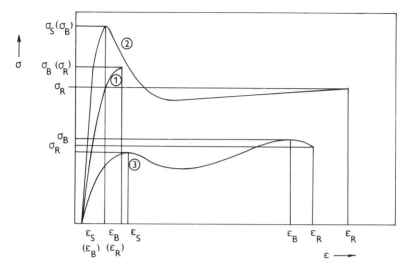

Fig. 2. Brittle material: 1, Plexiglas. Tough material: 2, Plexidur; 3, Poly-ethylene. Brittle materials described by the values: σ_R, ε_R, E_0. Tough materials described at least by the values: σ_R, ε_R, E_0, σ_S, ε_S. Rupture stress, σ_R; rupture strain, ε_R; yield stress, σ_S; yield strain, ε_S; peak stress, σ_B; strain at peak stress, ε_B; initial modulus, E_0.

tice. Brittle or tough behaviour may not always be the main property of a polymer: loading rate and the type of material can influence its character in a decisive way (Fig. 2).

When the load is removed slowly, the strain changes with stress in the same way as under loading, though the values of strain may be lower because of the influence of the destruction of chains or processes of crystallisation. If large strains (10–20%) have been reached, the course may differ quite considerably, with the elastic strains in reverse, perhaps with an anelastic component.

In inelastic deformation the greater part of the mechanical work is not converted into a reversible change of state, but is irreversibly used in overcoming internal friction. Owing to the low thermal conductivity, heat cannot flow away, and the temperature increases locally. This may even result in a point of transition to another state (softening). Consequently, cold forming is actually hot stretching, with the higher temperature zone gradually passing through the material and resulting in expansion of the region of flow.

Of course, this is not the only mechanism possible in the deformation of polymers: in orientated material which is already stretched, for example, a mechanism will appear that partly results in stress peaks being cut by further straining (analogous to the Bauschinger effect). At the same time, quantitatively different behaviour under qualitatively different straining, for example by compression and tension, is not precluded.

There is a difference of principle between the mechanical function of the structure of the model materials and that of the structural materials.

In polycrystalline materials, it becomes necessary to define the structure with respect to the distribution of impurities, grain boundaries, dislocations, bubbles, etc., which influence certain structurally sensitive properties, such as the elastic limit, the strength and the optical properties, whereas the so-called structurally insensitive properties, such as the elastic constants, are independent of such imperfections. From the point of view of similarity it will then suffice to consider only the identity or proportionality of the relevant material constants. Texture, arising from a preference in orientation of the structural elements, generally crystals, has not any parallel in model materials. A definite analogy can best be seen from the fact that during the plastic deformation of crystalline materials a certain orientation in the direction of straining is created, and that large deformation of linear polymers beyond the elastic limit leads to a certain unfolding of the macromolecules into a parallel straight-chain orientation. For example, in crystalline polyethylene one-directional tension causes grains to stretch into thin fibres, and the birefringence reaches high orders very close to white light; failure, then, sets in when these stretched fibres become disconnected and broken after being exposed to plastic deformation. Terylene and PTFE, however, which have similar structures, do not display local plastic deformation and their failure is brittle. Investigation of the biaxial orientation of polymethylmethacrylate has shown that such orientation does not exert much influence upon tensile strength and elasticity modulus.

From the nature of the deformation mechanism in polymers it follows that the permanent component of strain may be taken as partly plastic and partly viscous in nature. The ratio of both these contributions will vary for different materials, and modelling in plasticity will naturally be more effective with materials whose viscous deformation is less pronounced. Along the boundary between these groups lie organic materials, which exhibit both viscous and plastic deformation equally, e.g. wax, pitch and gelatine gel.[2-4]

Plastic deformation, which is a characteristic of metals, has no direct parallel in polymers. It can only be admitted in the sense of the given definition; the yield surfaces correspond to the region in the deformed material in which deformation is effected by normal stresses along surfaces, or lines, of least resistance to shear deformations from the moment of application of the state of stress. The state of stress varies during the course of deformation, this being approximately along isostatic, or isostatically related, surfaces or lines.

The diverse applications of polymers to modelling make it necessary to mention technically suitable ways of describing their time-dependent deformation, creep deformation and deformation by plastic flow. The general character of the creep deformation of materials also applies to model materials, although some specific peculiarities must be added owing to viscous changes occurring in the structure.

The double character of creep leads to the distinction of two types of macromolecular materials. On the one hand, there are those which deform with direct dependence upon time, that is by linear shear deformation which is described by the function $\gamma = \gamma(t)$, which the majority of model materials satisfy unless a certain limit of stress is exceeded. On the other hand, there are materials which deform non-linearly, as described by the function $\gamma = \gamma(\tau, t)$.

Linear viscoelastic behaviour is characteristic of both linear and crosslinked polymers, which are currently used as model materials. In pure, perfectly hardened polymers time-dependent changes are less pronounced; it is in plasticised polymers, or in normal polymers in the transition region, that pronounced viscoelastic behaviour becomes important.[26]

With greater forces and larger strains, often greater than 1%, caused by these forces, non-linear relationships have to be considered; in practice the composition of the total strain of type

$$\gamma(t, \sigma) = \gamma_1(\sigma) + \gamma_2(t, \sigma) + \gamma_3(t, \sigma) \tag{1}$$

proves useful. γ_1 is the reversible elastic strain, which can also be non-linear, γ_2 denotes the creep strain showing within a maximum of 500 h, for which Nutting's equation

$$\gamma_2(t, \sigma) = K\sigma^m(1 - e^{-qt})$$

can be used. The material constants K, m, q and the function γ_3 denote the time-dependent strain present after some 8000 h, after which time a constant rate of strain has been reached, which may be described by Ostwald de Waloo's law, in terms of the material constants B and m

$$\gamma_3(t, \sigma) = Bt\sigma^m$$

In plastic deformation time becomes important particularly by way of the strain rate. At different strain rates the mechanical constants of the model materials, and also their properties, change, so that the higher rates cause a certain work-hardening of the material; in order to produce the same strain quickly, greater stress has to be applied than for slower rates. The more rapid the deformation the more brittle the material becomes. Besides physico-chemical causes, the main reason why the mechanical characteristics of polymers change with rate of loading is to be found in the influence of heat in the course of deformation, which speeds up, or retards, the polymerisation of the monomeric remnants in the skeleton according to the extent to which displacements and deformations have been able to take place throughout the structure of those materials. A high loading rate does not allow plastic strains to develop and the material exhibits a higher modulus of elasticity and embrittlement. A low rate of loading allows the molecular chains to unfold and the displacements to propagate, the development of which with time constitutes creep strains, and also enables the material to acquire toughness. In practice this means that plasticity in polymers is directly linked to strain rate. The influence of strain rate thus assumes special importance in dynamic loading.

Theoretical considerations and experimental measurements have shown Poisson's ratio to be a quantity which, in practice, is independent of time in the glass or rubbery regions. In the range lying between these states Poisson's ratio is a time-dependent, steadily increasing function.

The assumption of non-compressibility is roughly acceptable for the rubbery state only, or in the case of large plastic strains when $\mu \rightarrow 0 \cdot 5$. Under dynamic loading, temperature and decreasing frequency cause Poisson's ratio to change.

The deformation behaviour of plastics makes it necessary to control, i.e. isolate, stress or suppress, the basic time-dependent mechanical conditions during the investigation. These are characterised by the various types of loading process, and it is thus necessary to select the loading process most appropriate for a particular investigation, bearing in mind the purpose of that investigation: constant or changing load, constant or changing strain, constant loading rate, constant rate of change of loading, constant strain rate and constant rate of change of strain.

With a view to simplifying the experimental procedure, loading may be controlled in two ways. The first entails rapid loading, blocking the strain, allowing a certain period of relaxation and then taking the measurement. This procedure is not generally applicable, since in model materials that are under constant strain, the stress in highly strained regions decreases more rapidly than in moderately strained regions, and the resulting stress–strain curves will be understandably different.

The second procedure entails maintaining a constant load during the experiment. In this case it would, however, be difficult to achieve large plastic deformations since loading must be known from the very beginning. Excessive loading would fail the model prematurely, and insufficient loading would stop the flow.

It is only possible to obtain the same stress–strain curves with negligible deviations for all points of the model as calibrated in a uniaxial test under conditions of constant strain rate. These conditions can be strictly maintained at only one point. Deviations from the expected shape of the stress–strain curve at other points are, however, almost insignificant in homogeneous states of stress and loading can easily be extended far into the plastic range.

In the inhomogeneous state of stress the strain rate varies both from place to place and also with time at each point, with the exception of the point at which the strain rate is maintained constant by control. Thus, each point of the model only corresponds to one valid stress–strain curve in the whole array of curves obtained at various strain rates. A certain instant of time in the experiment is represented by a certain point on each curve and the line connecting all these points is the stress–strain curve, the so-called resultant curve, valid for that instant of time. For the case when the stress–strain curves do not differ greatly, the resultant curve will, to a considerable degree, cover the whole deformation.

In the range of elastic and plastic strains model materials are practically homogeneous. In the range leading to failure, however, weaker points in the structure become significant. These include notches shaped into the model during manufacture, microcracks in the material, and places of imperfect polymerisation caused by ambient factors, such as temperature, humidity, etc. The strength limit, under otherwise identical mechanical conditions, may show considerable scatter and should be considered a statistical quantity. Imperfect homogeneity is also affected by the fact that during solidification or polymerisation a certain direction gains preference in the structure owing to the uneven distribution of the heat liberated.

In a broader sense the imperfect homogeneity and isotropy of plastics result in a scatter of the values of the mechanical and optical constants for different slabs of the same material, even though these may be from the same batch and the investigation may be concerned with different regions of the same slab. There is also the influence of the age of the material arising from final polymerisation of the remnants of the monomer over a long period of time, storage of the slabs, where the inclined or perpendicular position of the slabs predetermines the direction for long-term creep orientation, and other physicochemical causes.

The isotropy of the optical properties is limited by orientation birefringence caused by the manufacturing process manifesting itself in the direction normal to the thickness of the slab. Though thickness does not assert itself in the plane elastic states, it can influence the course of the plastic flow through spatial displacements.

3.2. Approaches to Inelastic Experiments

The condition that the strain rate must be maintained constant at every point of the solid in a photoplastic experiment can be circumvented with acceptable accuracy. In inhomogeneous states of stress, the stress may decrease at a certain phase of the process though strain tends to increase, which is not consistent with the assumption that stress must not be relieved during the process. The same measure of accuracy may also be achieved by a simpler process, in which the loading rate is the influencing factor.

In controlling the loading process through the loading rate, differences can, of course, arise between the rate of loading and the rate of growth of the initial strain as a result of creep or plastic flow. It is therefore necessary to limit the loading rate so that creep does not gain decisive importance and so that the two causes of growth of strain may be differentiated. A separate evaluation of the plastic strains by considering the effects of stress and the strain rate does not prevent the time superposition of both components in those cases where the strain rate acquires special significance.

It is obvious that although the structural and model materials may not be directly analogous, there are a sufficient number of parallel effects to ensure their applicability. It is necessary, however, to choose a material whose behaviour is most appropriate for the problem under consideration.

The physicomechanical properties of the model materials thus generally comprise a broad spectrum of characteristics: those investigated by measurements of tension and compression, bending and vibration, longitudinal vibration, flexural vibration, fatigue, and by means of ultrasonics, etc. Their full extent will have to be considered in the complex modelling of the similarity to the structural materials. As a general characteristic of the range of mechanical properties of plastics one can decide upon the relationship between damping and temperature, or between damping and frequency.

The application of theories of inelastic deformation to engineering problems is conditioned by the validity of the theories, either in general, or within certain limits, which necessitates a systematic evaluation of their phenomenology.

It is assumed that there exists a certain characteristic of the material which enables the unique determination of the state of stress at which plastic deformation begins and the effect of increasing loads is investigated under a certain ratio of the components of stress independently of the orientation of the principal axes of stress. The conclusions then verify particularly the assumption of isotropy, introduced by the criterion of plasticity as an invariant function by the stress components. In a similar way, a record is made of the influence of anisotropy; this can basically be threefold: crystallographic (due to texture, meaning preferred orientation of aggregated crystals); mechanical (the well-ordered arrangement of the different phases in the material, including pores, cracks, etc., arising from non-homogeneity); and lastly, deformative, or inelastic, due to the history of loading, as in the case of cold forming.

The most common way of characterising the deformation behaviour of a material is to determine the stress–strain curve for a suitable sample in a testing machine. From the shape of the curve, especially under cyclic loading and unloading, with the upper limit of loading being gradually raised, the proper allocation of the tested material to theory can be made. It is, of course, better to compare the theoretical results directly with tests carried out on a simple solid.

More detailed study leads to the investigation of the behaviour under a multiaxial state of stress. For a description and a graphical interpretation of these states, adequate expressions must be used, mostly of a relative nature. As a reference value, the stress σ_1, or the strain ε_1, is chosen for the case of uniaxial loading: the abscissae and ordinates are then specified in normalised units of $\varepsilon/\varepsilon_1$ and σ/σ_1. If the biaxial state is investigated, the loading programme is conducted so that the stress $\sigma = \alpha\sigma_1$ is kept constant; for the triaxial state $\sigma_2 = \sigma_3 = \alpha\sigma_1$. For every α we obtain a curve of the stress–strain diagram, and the deformation behaviour of the material is then described by this array of curves for both positive and negative values of α.

It is important to be aware that the stress–strain diagram is influenced by a number of factors. First of all it will vary with the type of the plotted strain and stress, and with the velocity of the loading force, of the strain rate. Bearing this in mind, and also neglecting the dead load of the sample, we realise that the stress–strain diagram does not represent the actual stresses and strains but the stress and strain increments produced by the external force. Further, every tested sample will have regions in which the state of stress will be three-dimensional, even if the sample is uniaxially strained. These regions are not taken into consideration despite the fact that owing to its symmetrical distribution the uniaxial state of stress is always the most favourable. There are also states of surface stress in a solid that are undefined, the intensity of which varies with the size of the body.

Finally, there is the influence of the elasticity and rigidity of the testing machine itself. The accuracy of the values read from the testing machine requires attention; it is thus necessary to check the geometry and kinematics of the mechanism of that equipment. In the most significant theoretical investigations, such attention to detail is of paramount importance.

4. INTERPRETATION OF DEFORMATIONS BY MEANS OF POLARISATION-OPTICAL EFFECTS

4.1. Physics of Birefringence

The electromagnetic nature of matter causes the molecules to create a field round their centre of gravity and, roughly speaking, the less symmetry the molecules display, the more pronounced are the aniso-tropic properties of the field. This anisotropy manifests itself in various ways and is especially marked in macromolecules, since their length exceeds their diameter by several orders of magnitude. The most pronounced optical result of this anisotropy is the refraction in two components of a ray of light passing through a mechanically stressed man-made material. This type of birefringence usually disappears when the load is removed, and it has therefore been labelled as temporary. The simplest way of observing this phenomenon is by means of visible light, using a transparent material, though observation of the phenomenon is in no way limited to these conditions only. In the case of non-transparent materials infrared radiation may be used, and in special cases X-radiation may also be employed. Wertheim's law describes the correlation between temporary birefringence and the elastic state of stress.

In addition most man-made materials exhibit double refraction up to their point of fracture. Physics, and especially physical chemistry, has so far provided few exact data concerning the basic properties and origins of the birefringent effect. Little is known of the dependence of this effect on the structure of the material and on its relationship to mechanical straining. Present knowledge of the nature of birefringence comprises basically some phenomenological observations and hypothetical considerations.

In a solid, mechanical forces cause atoms to move from positions of equilibrium; the distance between the centres of gravity, or nuclei, of the direct-bond atoms probably does not change, but the distances between the indirect-bond atoms, that is, those forming further branches of the basic skeleton structure or intramolecular bonds, and the valency angles, do change. In transparent man-made materials these changes result in a splitting of the light-beam into two linearly polarised mutually perpendicular rays as it propagates through the material, causing the electron shells of the atoms to vibrate. The phase of the original beam changes to two mutually different phase shifts in both components. Quantitative evaluation of the two components depends upon the so-called polarisability, which may be considered as an indication of the light transmissivity of the electron shell of the molecule. It is the determining factor in optical anisotropy that causes birefringence and has the properties of a symmetric tensor.

The light wave, having a periodically changing magnetic field, is capable of interpreting the polarisation of the molecule, that is, the quantitatively different behaviour in two, or three, directions, by the changing velocity of its propagation. The macroscopic behaviour of the substance, as one comprising molecules that possess this property, is then characterised by the refractive index n pertaining to a given direction. As the passage of light is to be considered two-dimensionally, the third dimension being the optical path, the magnitude of birefringence may be simply computed as the difference in the extreme indices of refraction in the two principal directions. If the distribution of molecular ellipsoids in the material is wholly random (this corresponds statically to the situation in liquids) then, macroscopically, the material exhibits optical isotropy. Anisotropy, as we have already indicated, arises in such a medium through the action of external forces that exert an influence on the existing state of equilibrium. Each orientation of the state of the structure, influenced by external forces, gives rise to the ellipsoid of polarisability for the macroscopic volumes of the material, with the principal axes of that ellipsoid aligning in the main direction of that orientation. It may be noted that such a cause will result in positive birefringence. The cause of the optical anisotropy may also be found in the density variations in different directions owing to the anisometric structure.

If we consider uniaxial tension, density decreases in the direction of the extension, and increases in the direction perpendicular to it; birefringence will be determined here with regard to the contribution from the direction of the principal and associate axes, not to the Wertheim constant. The anisotropy of density usually goes hand in hand with orientation anisotropy. In the resultant birefringence, these act in opposition and the sign of the resultant birefringence cannot be decided on the basis of present knowledge. Negative birefringence, however, seems to predominate.

During an isothermic process a macromolecular solid responds to external loading by the reaction Q (or by stress, if we relate the reaction to cross-section), which consists of two components

$$Q = \left(\frac{\partial F}{\partial l}\right)_{T,\eta} = \left(\frac{\partial U}{\partial l}\right)_{T,\eta} - T\left(\frac{\partial S}{\partial l}\right)_{T,\eta} \tag{2}$$

Here F denotes free energy, U the internal energy, S the entropy of the system, l the length, T the absolute temperature and η other internal parameters. The first component representing the length-dependent change of energy is the work done in changing the interatomic forces that retain the atoms in their equilibrium positions; it can be labelled energy elasticity. Since at that stage the weakest bonds have already failed, the change affects mainly the secondary valency bonds between the molecules, and their valency angles, which results in the anisotropy of polarisability. The principal valency bonds participate very little. The type of birefringence which corresponds to this mechanism is stress (or force) birefringence. It is obvious that this type of birefringence is, above all, a characteristic of low-molecular materials. In high-molecular materials experiencing large strain, the resulting value of birefringence receives a contribution from another component denoted with respect to its origin as orientation birefringence. Its mechanism is interpreted by the second term of eqn (2) which describes the length-dependent change in entropy and is denoted as entropic elasticity, though from the point of view of mechanics the term elasticity is a misnomer where restitution to the original state involves external physical intervention, as, for example, a change of temperature. It entails the orientation of the molecular chains (straightening) in the direction of straining. For kinetic, thermodynamic, statistical and mechanical reasons, however, this tends to revert back to a state of disorder. The value of stress birefringence δ_s will thus depend upon the distribution of the ellipsoid of polarisability of the chain atoms and upon the difference in the respective principal values of the resultant polarisability; this value is interpreted in terms of the basic law of photoelasticity, known as Wertheim's law

$$\delta_s = C\tau d(= C'(\sigma_1 - \sigma_2)d) \tag{3}$$

where C stands for the photoelastic constant, τ the maximum shear stress, and d the thickness of the medium in the direction of propagation of the light beam. Stress birefringence occurs in small, i.e. solely elastic, strains.

Orientation birefringence is caused by orientation, that is, unrolling and simultaneous straightening of the chains of molecules through the action of external forces. Measurable, or even better, separable, orientation birefringence appears in large strains only when free deformation can proceed unhindered, especially in fibres and foils. It represents a measure of the orientation of molecules or, in crystalline materials, of whole regions of crystal aggregates. Compared with stress birefringence, orientation birefringence is time-dependent so that it is capable of interpreting creep and relaxation effects as well. The sign of orientation birefringence is determined by the position of the axis of maximum polarisability with respect to the direction of the chains. If this coincides with the direction of the chains, as for example, in polyethylene, then the orientation birefringence is positive and its sign is the same as that of stress birefringence. If there are anisotropic structural groups in the molecule, for example laterally bonded benzol groups, the sign of orientation birefringence is reversed and for larger strains it may even overlap stress birefringence completely and bring about a change in the sign of the resultant birefringence, as in the case of polystyrene and polymethylmethacrylate.

The dependence of orientation birefringence, δ_o, upon the orientation of the macromolecules, characterised by strain, has been theoretically formulated by Kuhn and Grün for high-elastic, i.e. rubber-like, polymers. There is no linear relationship between birefringence and strain, except for negligible magnitudes. It is, however, possible to infer from these theoretical formulations direct proportionality between orientation birefringence and the stress σ for elastomers. This has the form

$$\delta_o = \frac{\bar{C}_o}{E}\,\sigma = C_o \sigma \tag{4}$$

It is interesting to note that the relationship between birefringence and strain is independent of temperature, whereas the dependence of the relationship between birefringence and stress upon temperature is inversely proportional; absolute temperature asserts itself in the function describing Young's modulus E.[2]

The general character of the orientation of macromolecules in the structure due to loading also reveals stretching in polymethylmethacrylate.[7] Figure 3 shows measured values for the orientation angles and Fig. 4 the linear dependence of birefringence on Herman's orientation function F_{or}.

The two types of birefringence always appear simultaneously in

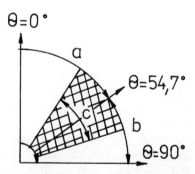

Fig. 3. Mean angle of orientation θ of the segments during uniaxial and biaxial stretching. (a) Uniaxial orientation; (b) biaxial orientation; (c) achievable in practice. $\theta = 0°$ for ideal uniaxial orientation, $\theta = 54\cdot7°$ for isotropy, $\theta = 90°$ for ideal planar orientation.

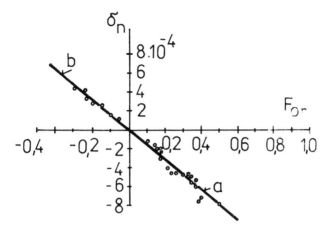

Fig. 4. Birefringence of stretched Plexiglas GS 209 as a function of molecular orientation. (a) Uniaxially stretched; (b) biaxially stretched.

man-made model materials, though with different magnitudes, and owing to the nature of their mechanism they exhibit different properties. Stress birefringence can exist almost solely in solid polymers in the glassy state; beyond the temperature of the glassy state its presence is perhaps insignificant. Orientation birefringence characterises, basically, non-linear behaviour and accounts for time-dependent changes in the resulting value of birefringence. The two types of birefringence can in principle undergo relaxation, but the rates of relaxation corresponding to a given temperature are considerably different. Stress birefringence is not influenced by temperature very much, whereas orientation birefringence is extremely sensitive to temperature variation. It follows from physicochemical investigation that the composition of the resultant birefringence is an addition of the two types. The final value of birefringence, denoted as deformation birefringence, in order to distinguish it from the two components, is thus

$$\delta = \delta_s + \delta_o \tag{5}$$

There is an hypothesis that orientation changes both the modulus of elasticity of the material and the photoelastic constant in Wertheim's law. This assumption has been tested on materials which display orientation birefringence (arising from the technology of their production), even before mechanical loading.

Rheological studies, too, refer to the relationship between birefringence and the deformation mechanism.[8] From an investigation of relaxation behaviour it follows that the time-dependent change in total birefringence is the result of three processes, namely, a decrease in stress birefringence, an increase in orientation birefringence resulting from a reduction in stress, and a recession of orientation birefringence, disorientation, caused by the relaxation of entropic stress. In terms of mechanics, the relaxation of the energy-elastic stress results in entropy-elastic stress, which also relaxes, though much more slowly, owing to the fact that its magnitude is about 1000 times smaller; the entropy stress starts relaxing only after the energy stress has nearly died away. This indicates that in plastic strains the influence of entropy-elastic stress is negligible. It is only over a very small range, in which the relationship between stress and birefringence is exactly linear, that the relaxation rates of both these quantities are the same. Neither kind of stress is thus more important to deformation, but the one develops as a consequence of the other. Relaxation of stress is a thermodynamically irreversible process and, at constant temperature in the glassy range, it is mechanically irreversible, too. Heating causes reversibility of the residual strain, thus denoted as quasiresidual, which is a phenomenon of thermal recovery. Both kinds of stress, however, manifest themselves by birefringences of comparable intensity and that is why, especially in the measurement of time-dependent change in the resultant birefringence, the agreement, or disagreement, of the signs of the stress and the orientation birefringence is of great importance.

The addition of a plasticiser, manifesting itself by a lowering of the modulus of elasticity, by the time-dependence and by a drop in the temperature of the transient state, has a varying influence upon birefringence according to the way in which the plasticiser is bonded in the structure. If there are no secondary bonds strong enough to tie it to a final position, its influence upon orientation birefringence is zero. In stress birefringence the plasticiser will contribute to the resultant value as if it were autonomous in the pure glassy state, irrespective of whether the plasticiser can be brought to the glassy state or not.

Some studies assume the birefringence to be composed of the crystalline and amorphous contributions to the stress-optical retardation spectrum:[9] this idea corresponds with the deformation changes in the structure of linear polymers, mentioned above. The separation of the two contributions has been made on the basis of the two-phase hypothesis for low-density polyethylene.[8]

4.2. Residual Birefringence

In the deformation of polymers, the influence of initial stresses is also of importance.

In higher polymers these are mainly stresses originating in the solidification of the orientated structure during transition to the glassy state, and also stresses remaining in the structure after removal of the large load responsible for the initiation of plastic flow. If the material has not been subjected to a further thermomechanical process, we have the case of so-called 'stress-freeze', that is, freezing of the entropy-elastic stresses in an orientated structure. Heating such a structure up to the transient temperature, or higher, will result in disorientation due to thermal reactivity based on micro-brownian motion, that is, the displacement of segments of the chains of molecules resulting in a statistically random, disordered structure; the stress will vanish, and the material will undergo shrinkage.

If, during heating, changes in length are prevented, as in relaxation, disorientation sets in, based on macro-brownian motion, which comprises micro-brownian motion with the possibility of whole chains slipping over one another, and the stress will vanish, too. Though thermodynamically irreversible, shrinkage is a mechanically reversible process, since hot stretching will restore the original length. Disorientation with a restriction of length variation is, on the other hand, a thermodynamically and mechanically irreversible process, since the given deformation can in no way be changed by further thermal processes.

The so-called edge effect does not indicate internal stress in the sense described above; it is caused by physico-chemical changes in the surface layers of the material. We are concerned here with a migration of components with low molecular weight from the material into the surrounding medium, or with the absorption of humidity from the environment. In the structure of a solid polymer, pores exist whose diameters are comparable to the size of the molecules of gases or liquid substances that constitute the environment. Such pores, however, can also be filled by the remnants of the monomer that have not participated in polymerisation and which then diffuse into the environment. Thus a certain re-grouping, loosening or concentration, of the already fixed macromolecules takes place leading to changes in the polarisability of the macromolecules, and thus also to birefringence. The process can proceed as an increase or decrease in the weight of the material. The migration of particles from the structure into the environment is, of course, an irreversible process; this effect cannot be removed by annealing. With increasing distance from the surface the intensity of birefringence grows less, since the process penetrates into the depth of the material with decreasing velocity.

4.3. Dispersion of Birefringence

The birefringence of model materials displays yet another physical phenomenon, that of chromatic dispersion. By dispersion of birefringence is understood the dependence of the relative phase shift upon the frequency or wavelength of the incident radiation.

Mönch has pointed to the relationship between the increasing dispersion of birefringence in celluloid, which in the elastic range amounts to 3–8%, and the increasing plastic strain, with dispersion reaching 72%. Furthermore, he has used this to obtain a measure of the magnitude of the plastic strain.[15]

It seems advisable to consider dispersion of birefringence as a relative quantity using the formula

$$D=100\frac{(\delta\lambda)_{Na}-(\delta\lambda)_{Hg}}{(\delta\lambda)_{Na}}(\%) \tag{6}$$

which expresses the percentage rate as a ratio of the difference in the products of birefringence δ and the wavelength λ for sodium light (5896 Å) and blue mercury light (4360 Å) to the value of that product for sodium.

A marked difference between the dispersion of birefringence in the elastic range and that in the inelastic range has been observed for celluloid. In other materials, for example, polyester resins, or CR-39, dispersion is independent of the elastic or inelastic strain. These materials are densely or totally crosslinked, whereas in linear celluloid the structural elements can move more easily, aided also by the presence of camphor. A considerable change in the magnitude of dispersion must result from permanent changes originating in the structure through plastic deformation, so that this phenomenon is indicative of the structural influence upon optical properties of polymers under mechanical straining. It thus serves as a measure of the magnitude of orientation birefringence.

Dispersion must also be considered when singular points are investigated using different wavelengths. In the elastic state of stress it already causes a shift of the singular point, and thus also a change in the position of isoclinics, as mentioned by Vandaele-Dosche and Van

Geen who, using wavelengths in the range 365–700 nm, determined dispersion in epoxide to be 7–8%.

4.4. Properties of Model Materials

Deformation and optical properties of the main types of specific model materials yield information about the applicability of these materials. One can find a complete survey of the necessary parameters in ref. 1 or in works dealing with individual materials (e.g. refs 10 and 11). Considering the specialised nature of this book only those parameters important in present day photoplastic and photoviscoelastic mechanical investigations are presented here.

Because the analogy between plastic strain and movements in polymer structure is due to singularities in their geometrical representations, the fundamental point is the shape of the macroscopic yield criterion (as well as the volume changes related to it). For reasons of mathematical simplicity the yield criterion of Huber–Hencky–Mises is given preference. With practically the same degree of accuracy for the range of interest, the Tresca criterion can just as easily be used and it is up to the researcher which one is applied.

Table 1 gives a survey of those properties of model materials. Values can differ from measured ones due to a different manufacturer, age, rate of loading or surrounding conditions. Note that only celluloid and polycarbonates have a clean-cut yield point. Celluloid or polycarbonates produced by extrusion are inapplicable due to inherent initial birefringence and a fixed structural arrangement. Celluloid and unsaturated polyester are endowed with considerable chromatic dispersion. With polystyrene and unsaturated polyester the sign for orientation birefringence is opposite to that for stress birefringence. Derivatives of cellulose have a linear strain-optical relationship and the birefringence is apparently independent of the velocity of strain. They are materials which were originally powders, and plate and block shapes are produced by melting at higher temperatures and pressures. An outstanding property of these new materials is their ability to scatter light, so that three-dimensional problems can be solved using the Tyndall effect. Detailed investigation of polycarbonate has shown that inelastic deformation accompanied by large strains change its structure to such an extent that it loses its original character and becomes, virtually, a different material; thus the elastic-plastic states cannot be evaluated, generally, in the plastic region.

TABLE 1
Basic Properties of Model Materials

Material	Limit of proportionality (MPa)	Maximum elastic strain (%)	Young's modulus (MPa × 10³)	Ultimate strength (MPa)	Poisson's ratio, elastic/plastic range	Optical constant (Na) (MPa. fringe. cm)	Creep	Residual birefringence	Application[a]
Polyethylene— high pressure	6	10	0·06	30	0·35/0·41	0·11	Slight	Yes	PhP, VE
Polyethylene— low pressure	~20	20	0·35–0·70	25	0·35/0·41	0·11	Slight	Yes	PhP, VE
Celluloid	+17, −18	0·7, 0·5	2·0–2·8	39	0·35/0·45	+3·9, −4·2, considerable dispersion	Perceptible	Yes	PhP
Cellulose acetate butyrate			0·6			0·3–0·75	Slight	Yes	PhP
Cellulose propionate	+14	1·6	0·9		0·4	2·5	Slight	Yes	PhP
Polystyrene	+15, −55	0·5, 1·9	3·0	+30, −80	0·35	6·3	Very slight	Turbid	VE, PhP
Polycarbonate	34	6	2·3	62	0·43	0·75	Slight	Yes	VE (PhP)
CR 39	18	0·55	2·8	+37, −46	0·33	1·85	Perceptible	Yes	VE
Polyester, unsaturated	5	2·5	1·9	20	0·35–0·40	2·0, strong dispersion	Considerable	None	VE
PVC	20	0·7	2·9	60	0·37	5·7	Slight	Yes	VE

[a] PhP, for photoplastic measurements. VE, for viscoelastic measurements.

The above characteristics of model materials are considered for applications at room temperatures. Sometimes it is more appropriate to increase the temperature to get yielding or flow behaviour.

5. THEORIES OF THE INELASTIC BIREFRINGENT EFFECT

5.1. Attempts to Generalise the Polarisation-Optical Effect

The complexity of birefringence phenomena on the one hand and the far-from-ideal properties of the material medium on the other necessitate making use of, or even giving preference to, phenomenological theories, for they can be relied on to supply the required data with satisfactory accuracy and give a one-to-one correspondence of the suitable (control) quantities, although not all details of the process may receive full physical representation.

Research into the viscoelastic behaviour of plastics, and the development of the theory of viscoelasticity, have led to a number of proposals which attempt to relate that type of deformation to the phenomena of birefringence, and to tests of the applicability of these relationships in practice.[10,12]

The deformation properties of model materials do not exhibit such a high degree of similarity that they could be described satisfactorily by a single equation; the possibility of describing them becomes even smaller if the optical effect is also considered. Theories relating to a single kind of material only, or those concentrating on only one definite interval of stress or time abstracted from tests carried out on a number of related materials, are therefore preferred, since theories of this kind are, generally, those most effectively put to practical use.

An attempt to incorporate the influence of macromolecular structure upon the stress-optical effect, strictly in the elastic range of the ideal amorphous homogeneous polymers, is represented by the work of Antropius and Lind,[13] who view the phenomenon as a superposition of several physical effects in a mechanically loaded model material, viz. the anisotropy of the distribution of polarisable particles, deformation anisotropy (distribution of chemical bonds) and the variation in the lengths of chemical bonds. These effects manifest themselves generally in the values of anisotropic refractivity and in the generalised Lorentz–Lorenz equation for a deformed medium; consideration of these facts leads to variations in the values of the main indices of refraction as functions of strain and to resultant photoelastic constants. Despite the fact that this theory, due to its physical background, does not depart from the confines of electro-kinetic theories, its inclusion of the parameters of the structure of the solid stimulates further interest. Coker and Filon were already aware of the limitation of the validity of Wertheim's law to perfectly elastic materials and then only to the linear range of their deformation. They were the first to suggest an equation for birefringence in the non-linear deformation (creep) range which embraces both stress and strain

$$\delta = a(\sigma_1 - \sigma_2) + b(\varepsilon_1 - \varepsilon_2) \tag{7}$$

where a and b are material constants for a given thickness of the model. This type of relationship between the quantities involved has persisted until now; the describing function, however, changes according to the needs of the author.

It was not until Mindlin attempted to describe viscoelastic behaviour by means of the four-member model that the theory of the photomechanical method of inelastic states reached the starting point of its new development.

In most cases the inelastic strains are evaluated from the measurement of birefringence with the aid of gauge curves.

Mönch assumed the criterion of plasticity for celluloid to be the constant octahedral shear stress.[14] He considered birefringence to be dependent not solely upon stress and strain but upon the ratio of the two principal stresses σ_2/σ_1. He therefore recommended using the value of dispersion of birefringence, assumed to be a measure of the intensity of the octahedral shear strain, as the criterion for finding the boundaries of the plastic region and as a measurable birefringent effect.[15] Netrebko has stated that a non-linear increment in the difference in the values of birefringence, belonging to two different wavelengths and measured at rising stress levels, is proportional to dispersion in the plastic range.[16]

Frocht, who was particularly interested in research into isoclinic lines, considered birefringence to be related to stress only. He excluded the influence of time by letting creep develop almost fully during 240 min. He then stated birefringence to be uniquely dependent upon the differences of the principal stresses up to 8% irrespective of the type of state of stress. In his investigation into the cause of the change in direction of stress during loading in a three-dimensional state of stress, he concluded that the history of individual stresses had no effect whatever on the resulting birefringence.[17]

Ohashi proceeds in his photorheological method from a mathematical description of the rheological deformation of viscoelastic materials with the relevant material constants included, and is able to determine the time-values of strains, which appear in the expression for birefringence. In his method,[18] the analysis of plastic states is made solely on the basis of the analogy between time-dependent deformation and plastic deformation, which follows from the similarity of the stress–strain curves. The relationship between birefringence and stress is then formulated so as to reflect the experimentally determined relationship. It was found that the ratio of the principal stresses did not, in practice, influence the mechanical and optical constants and that its effect is negligible. If, however, the result was to be made more accurate, material constants calibrated for the appropriate stress ratio would have to be used. It was further found that during unloading the constants were both independent of this ratio and of the stress rate applied before unloading. Calibrating the mechanical and optical constants involves a number of tests yielding values for various stress rates, or for the case of the principal directions of stresses moving in rotation. In practice it was realised that calibration carried out on a uniaxially loaded specimen was sufficient. The method presupposes knowledge of 17 constants, and thus requires a computer to process the data. It yields values for stress, and strains have to be determined separately, but it proceeds in a way which prevents the experiment fulfilling its main objective, that of furnishing more complete information on the process under investigation than the accepted analytical description.

5.2. Theory Based on the Double Nature of Birefringence

The physico-chemical conclusion concerning the double nature of birefringence, namely the deformation and orientation components, is a relationship between the optical effect and the mechanical state.[1]

Experiments carried out on celluloid have shown that stress birefringence depends upon stress, since it fully manifests itself after instantaneous and continuous loading, at time $t = 0$, when creep changes are out of the question.

Stress birefringence δ_s has been expressed by Brewster–Wertheim's law under zero initial conditions

$$\delta_s = C_s \sigma d \tag{8}$$

where the value of the photoelastic constant C_s may in some materials depend upon circumferential conditions.

The value of C_s for a given material differs only very slightly from the average photoelastic constant C, so that C_s would have to be calibrated only in extremely precise measurements and then used for evaluation at individual points.

The magnitude of the orientation birefringence is affected by the strain rate in the same way as by time in general through the magnitude and the gradient of strain. Thus, if the function $\varepsilon = \varepsilon(t)$ is known, orientation birefringence can be related only to strain, with the exception of the initial states of birefringence, since its value also depends upon the rate and acceleration of strain. From the point of view of practical use we can limit ourselves to the assumption of the initial state with zero state of stress and strain (and without initial birefringence of any origin) and thus control the loading process in such a way as to let the time exert its influence through deformation only. The influence of time can then be investigated independently of loading.

For the range of inelastic strains ε_p and $t = $ constant, we can then formulate the linear dependence

$$\delta_o = K_o(\varepsilon - \varepsilon_p)d \tag{9}$$

where K_o is an optical constant similar to that occurring in Wertheim's law.

Besides δ_s and σ_o, time exerts the greatest influence on the final birefringence. The influence of time is exerted either directly in creep effects, or through the rate of strain and its acceleration. The boundary between slow loading and instantaneous loading, or at least loading proceeding at a rate that prohibits creep changes to show, is determined by the minimum effective rate of deformation, i.e. that rate which results in plastic deformation. At higher rates of loading the material acquires elastic properties up to the point of rupture: the dependence of birefringence upon stress is described by Wertheim's law. At lower rates of loading the material displays creep deformation, causing increased birefringence corresponding to the increased stress applied.

If a model material without a clearly defined yield point is used, both types of birefringence develop from the very beginning of static creep loading. If constant loading continues, creep deformation and the corresponding changes in birefringence appear at every point.

Considering the conclusions noted above, the resulting birefringence can be described by the equation

$$\delta = \delta_s + \delta_o \qquad (10)$$

the stress and the orientation birefringence being governed by the laws formulated by eqns (8) and (9). By substitution into eqn (10) we obtain an equation which at first does not seem to possess any physical meaning because it expresses strain and stress as a sum, each associated with a different constant. A relationship of the type $\delta = \alpha\sigma + \beta\varepsilon$ may be quite common but it is upon physical grounds alone that they are usually contested.

If celluloid, or some other material, is proportionally loaded, the relationship between birefringence and stress acquires a form that corresponds to the graph in Fig. 5; its principal sections are straight lines. Of course this graph also represents the situation after constant loading for a duration of time t. The value of the constant C, pertaining to the range of the elastic state of stress, will be the value corresponding to the time t obtained from the calibration test.

Stress is, in physical terms, true stress, not conventional stress, acting at the point under examination. The linear dependence of δ_o upon the

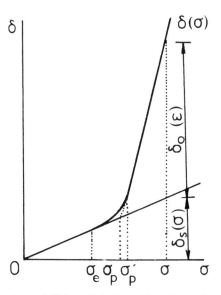

Fig. 5. The graph of eqn (10) in celluloid and in other polymers. Elastic range $\langle 0, \sigma_e \rangle$, transition range $\langle \sigma_e, \sigma_p' \rangle$, plastic range $\langle \sigma_p, \sigma \rangle$.

inelastic strain is valid not only for celluloid under gradual loading or creep, but also for CR-39 and for the epoxides, where the linear dependence $\delta - \sigma$ is of creep origin. Equation (10) thus achieves its real content, with ε_p being the maximum elastic strain,

$$\delta = [C\sigma + K(\varepsilon - \varepsilon_p)]d \tag{11}$$

For the reasons given above the form of the equation also acquires a physical meaning as the sum of the two components of birefringence of different origin. The constant C stands here for the photoelastic constant of Wertheim's equation and if birefringence is expressed in multiples of the wavelength of light used, and if strain is expressed as a percentage, the constant K (cm^{-1}) is determined by calibration tests carried out for instantaneous loading and for constant time. Equation (11) justifies its status as a physical relationship between birefringence and the mechanical state beyond the elastic limit in linearly birefringent polymers.

The relationship between birefringence and mechanical state retains its physical interpretation in the biaxial state of stress as well, so that

$$\delta = \delta(\tau) + \delta(\gamma) \tag{12}$$

and

$$\delta = 2[C\tau + K(\gamma - \gamma_p)]d \tag{13}$$

C again denotes the photoelastic constant, K, the photoplastic constant, having units cm^{-1}, γ the magnitude of the total shear strain and γ_p is the maximum elastic strain up to the point where the participation of non-linear strain in the magnitude of birefringence is negligible. The determination of the constant C is well known from photoelasticity; the constant K is determined by calibration tests at the strain rate that corresponds to the loading process.

Doyle[19] has recently attempted to unify the theories based on the birefringence effect in photomechanical methods using, first, Maxwell's equations to establish certain results independent of material properties and, second, proposing a model of polymer behaviour that essentially dissociates optical properties from mechanical properties. His optical constitutive equation reads

$$\frac{\delta\lambda}{d}\begin{Bmatrix}\cos 2\theta\\\sin 2\theta\end{Bmatrix} = \frac{1}{2n}\begin{Bmatrix}e_x - e_y\\2e_{xy}\end{Bmatrix} = C_\sigma\begin{Bmatrix}\sigma_x - \sigma_y\\2\sigma_{xy}\end{Bmatrix} + C_\varepsilon\begin{Bmatrix}\varepsilon_x - \varepsilon_y\\2\varepsilon_{xy}\end{Bmatrix}^e + C_p\begin{Bmatrix}\varepsilon_x - \varepsilon_y\\2\varepsilon_{xy}\end{Bmatrix}$$

$$\tag{14}$$

where n is the average index of refraction of the medium, d is the thickness of the model, δ is the isochromatic fringe order, θ the principal direction of the dielectric tensor (optical isoclinic angle), e_i the components of the dielectric tensor, C_σ the experimentally determined stress-optic coefficient and C_ε and C_p are strain-optic coefficients, for recoverable and permanent strains, respectively, subscripts ε and p referring to the recoverable and permanent strains, respectively.

In elastic-plastic material eqn (14) reduces to

$$\frac{\delta}{d}\begin{Bmatrix}\cos 2\theta \\ \sin 2\theta\end{Bmatrix}=\frac{1}{\lambda}\,K^*\begin{Bmatrix}\sigma_x-\sigma_y \\ 2\sigma_{xy}\end{Bmatrix}+\frac{1}{\lambda}\,C_p\begin{Bmatrix}\varepsilon_x-\varepsilon_y \\ 2\varepsilon_{xy}\end{Bmatrix} \tag{15}$$

with

$$K^*=C_\sigma+\frac{1+\mu}{E}(C_\varepsilon-C_p)$$

Assuming the mechanical constitutive equation for proportional loading,

$$\begin{Bmatrix}\varepsilon_x-\varepsilon_y \\ 2\varepsilon_{xy}\end{Bmatrix}=\left[\frac{1+\mu}{E}+\Phi_{\text{mech}}\right]\begin{Bmatrix}\sigma_x-\sigma_y \\ 2\sigma_{xy}\end{Bmatrix} \tag{16}$$

where Φ_{mech} is a non-linear function of stress invariants, we obtain

$$\frac{\delta}{d}\begin{Bmatrix}\cos 2\theta \\ \sin 2\theta\end{Bmatrix}=\left[\frac{1}{f_\sigma}+\frac{1}{f_\varepsilon}\,\Phi_{\text{opt}}\right]\begin{Bmatrix}\sigma_x-\sigma_y \\ 2\sigma_{xy}\end{Bmatrix} \tag{17}$$

(with fringe constants f_σ, f_ε), where $\Phi_{\text{mech}}=\Phi_{\text{opt}}$, which leads to the linear relationship between birefringence and Φ_{mech} for uniaxial stress within the whole interval of loading. This theory comprehends the notion that the isochromatics do not in general correspond with the principal stress and strain differences and that the isoclinics do not follow the principal directions of stress or strain, so that the shear-difference method cannot be applied for the evaluation of mechanical state.

Unfortunately this theory has not been corroborated for time-dependent effects, where the validity of eqn (14) is uncertain.

The value of this theory lies in its confirmation of the direct relationship between mechanical state and the dielectric field in the material structure; it disregards, however, the fact that the necessary determination of the functions Φ_{mech} and Φ_{opt} prior to the investigation of the model makes it of limited applicability.

6. SEMI-PHOTOPLASTIC PROCEDURES

If attention is paid only to one aspect of plastic deformation, viz. its final magnitude, as a basis for analogue solutions, one only considers the semi-photoplastic procedure. This is not totally correct and is only acceptable in special cases. For example, inelastic strains in a three-dimensional body were investigated with an epoxide resin model just below the transition temperature.[20] Plastic yielding of metal specimens under hot forming was simulated with the help of highly strained models of softened epoxide at increased temperature.[21] Considering the complexity of the behaviour of the structure of a hot-rolled product and the lack of any other adequate method, this approach is justified and the implanted error is substantially smaller than those incorporated in other approaches.

7. INVESTIGATION OF PHOTOVISCOELASTICITY

The problems here concentrate on a mathematical description of the time-dependent relationship (providing only elastic or viscous strains are considered), and not on the nature of the phenomenon. Considering irreversible deformations, the problems coincide with those of photoplasticity. That is why methods of optical polarisation of reversible deformation, or of viscous flow, have been elaborated with more accuracy than photoplasticity in its present state would allow.

Since the relationship between optical and mechanical quantities has attracted the attention of physical chemists researching into polymers, a large number of methods have been developed making use of the birefringent effect, dichroism, light scatter, X-rays, etc., and constituting an independent branch known as rheo-optics. As a method, when using polymers as modelling materials, the optical polarising method must take account of the viscoelastic behaviour of these materials. The viscoelastic phenomenon can itself be taken advantage of in the solution of problems of elastic strain with the time component. It is not possible to add anything to photoviscoelastic theory and practice in the form set forth by Coleman and Dill[22] or in a more complex form by Rivlin and Smith.[23] Thus, we only meet with those specialised applications of model solutions where the methodology (e.g. in combination with holography) and properties of the material are of practical importance.

The theory of photoviscoelasticity proceeds from the assumption that a relationship similar to that existing between time or temperature, and mechanical behaviour, can also be detected between optical and mechanical effects. This hypothesis is based upon the view held by physical chemists researching into polymers, namely, that the relationships formulated in this way need not necessarily depend upon the microstructure. This virtually means neglecting the dependence of the optical constant upon stress or strain, which has not been fully elucidated as yet, but which according to experiment does not seem to be of decisive importance. Investigations have mainly been carried out on crosslinked polymers, which exhibit a rather less pronounced viscoelastic behaviour in the glass and rubber range than in the transition range, in which loosening of the bonds predominates, and not some rearrangement of the inner configuration.[24] The optical constant is usually dependent upon strain rate, especially in polymers with a low elastic modulus; on the other hand, certain materials, e.g. CR-39, maintain a constant average value of optical sensitivity even at different rates.

From the point of view of stress analysis, the difference between elastic and viscoelastic deformation is to be sought in the mechanical and optical behaviour of the material as represented by the relationship between stress, strain and birefringence. This analysis presupposes a complete list of the mechanical and optical characteristics as functions of the rate of strain and of temperature. The appropriate values of C_σ and σ_e can then be substituted into equations according to the conditions under which the experiment is conducted.

A rather complicated situation arises with the non-linear viscoelastic behaviour of model materials. Simplification based on the principal quasi-linear and principal cubic theories then applies. The correspondence between mechanical behaviour and the optical effect is based on the similarity of related curves for the usual model materials or specially prepared polymer mixtures or copolymers.[25]

The mechanico-optical relationships under dynamic loads are still not properly understood. Viscoelastic behaviour and the related optical response for some selected materials is the object of contemporary investigations.[26]

The evaluation of viscoelastic measurements involves consideration of the coincidence of principal directions. In an isotropic, homogeneous viscoelastic medium, where the coefficients of the basic relationship are linear operators, special conditions must be applied, so that coincidence of all the axes may be ensured beforehand. In photovis-coelastic materials, however, where relationships of this kind hold for optical characteristics as well, the direction of the principal axes, and the differences of the principal stresses, can be determined through measurement of the isoclinics and isochromatics as functions of time. Thus, in general, sets of isoclinics pertaining to different times form an essential part of the data of photoviscoelastic measurement; measured values of the differences in normal stresses in a given system of coordinate axes and parameters of isoclinics make it possible to determine the differences in the principal stresses and the correspond-ing directions of the principal axes.

As has already been stated, in the theory of viscoelasticity the behaviour of linear materials shows a similarity between the depen-dence of mechanical quantities upon time and the dependence of these quantities upon temperature. An analogous dependence can be found between the optical coefficient and time and temperature.

The principle of the time and temperature shift is applicable for dynamic reasons to both the mechanical and optical properties of polymers.

Analogously to the linear theory of viscoelasticity, the function for birefringence can be written

$$\delta(t) = C_o \sigma(t) + \int_0^t \Delta C(t - \tau) \frac{d\sigma(\tau)}{dt} d\tau \qquad (18)$$

where C_o stands for initial stress-optical coefficient and $C(t)$, separated into two components, is written as

$$C(t) = C_o + \Delta C(t) \qquad (19)$$

It is obvious that the parameters on the right-hand side of the equation have no connection with the physical nature of birefringence. Tougui *et al.*[12] proceeded from this representation to develop an appropriate relationship for non-linear photoelasticity and derived the expression

$$\delta(t) = q_0 C_o \sigma(t) + q_1 \int_0^t \Delta C(\psi - \psi') \frac{d}{d\tau} [q_2 \sigma(\tau)] d\tau \qquad (20)$$

in which the quantities

$$\psi(t) = \int_0^\tau \frac{d\tau}{\alpha_\sigma}; \qquad \psi'(\tau) = \int_0^t \frac{dt}{\alpha_\sigma}$$

and q_0, q_1 and q_2 are time-independent functions of stress level. The function α_σ may be both stress and time dependent while the quantities q_0, q_1, q_2 and α_σ have to be evaluated from calibration tests. It follows that for $q_0 = q_1 = q_2 = \alpha_\sigma = 1$ linear theory is recovered. Equation (20) is thermodynamically consistent, involves a mathematical statement of the time–stress superposition principle and is analogous to the time–temperature superposition principle. It is applicable (at least) for the polycarbonate used in the study.

A special region of research on time-dependent deformations involves rheological problems, especially the problem of creep. Here, time can cause plastic deformations to appear and these then enter the photoplastic solution proper. But modelling mostly concerns the creep of structural materials, where in intervals practically anticipated strains do not reach such magnitudes and the problem becomes viscoelastic in nature. The experimenter is then more interested in the method of measurement and in the simulation of the structural material by an adequate polymer. Most of the cited work is of this type, e.g. that dealing with creep in concrete or with viscoelastic effects in the vicinity of crack roots.

Rheo-optics, as such approaches may, in general, be labelled, is an extensive field of research, dealing with relationships between birefringence (and other optical effects) and the structure of polymers. In rheo-optics the electromagneto-optical theory of birefringence assumes special importance as the direct interpreter of the electromagnetic field of molecules and its elaboration becomes a strong driving force in the development of the method.

The optical polarising method of time-dependent inelastic strains is applied to problems that do not require any special or new physical theory. Use is often made of the theory of graphic correspondence and evaluation is based on analytic geometry, apparently owing to the physical triviality of this approach. Due to the nature of the birefringent effect the most appropriate approach seems to be the theory of the double nature of birefringence.[27] Since creep does not involve complex states of stress, the application of this method to rheological problems is very promising because of its simple interpretation of stress and of the elastic and inelastic strains.

8. CONTROLLING THE EXPERIMENT

8.1. Evaluation of Measurements

Analysis of the relationship between isoclinics and the principal directions of stress and strain is an important part of the inquiry into the relationship between mechanical quantities and birefringence. The main point of the problem is whether—specifically in celluloid—with the increase in optical anisotropy, mechanical anisotropy must appear, as a consequence of the superposition of two different plastic states. Investigations by Frocht[17] have shown that mechanical coincidence exists between the directions of principal stresses and those of principal strains, provided the direction of the principal stresses does not change in the course of loading. If, however, new loading is superimposed upon the prior plastic state of stress, and the directions of the principal stresses in the new state differ from those in the preceding one, the resultant principal stresses in celluloid take directions which do not coincide with those of the principal strains. It is worthy of note that the same phenomenon can also be observed in metals, for example, aluminium and steel, which puts celluloid in the advantageous position of a material suitable for investigations of this kind.

In connection with the problem of mechanical coincidence, optical coincidence was also studied. The latter exists in principle only between the directions of the principal stresses and the parameters of the isoclinics. The coincidence between the directions of the isoclinics and the resultant principal stresses is not affected by their rotation; neither is the magnitude of birefringence. This fact is of decisive importance for the evaluation of isostatic lines of elastoplastic states of stress. The isostatic lines, drawn on the basis of isoclinics, as in photoelasticity, always represent the directions of principal stresses; at the same time, they represent the directions of principal strains only in cases of mechanical coincidence. If the principal directions of mechanical quantities do not coincide, the isostatics will not yield information about the displacements arising in the plastic region, and the system of the orthogonal shear trajectories cannot be regarded as decisive for the determination of yielding and slip lines in the body under investigation.

In photoplasticity, the meaning of isochromatics as loci of points exhibiting the same value of maximum shear stress, besides their meaning as loci of points exhibiting an equal value of birefringence, is connected with the behaviour of model materials in shear, or with the difference between straining in tension and in compression. The anisotropy of the properties of celluloid in tension and in compression is a factor which can complicate the evaluation of plastic states of strain in models made of that material. Those expressions normally valid can no longer be made proportional and in the expression $\tau = \alpha\gamma$, α must be defined for every difference $(\sigma_i - \sigma_j)$ or $(\varepsilon_i - \varepsilon_j)$.

The dispersion of birefringence in plastic deformations is proportional to the larger of the two principal strains. The same condition, however, is also decisive for the onset of the plastic state in the plane state of stress. The time-dependent change in, dispersion indicates dependence upon strain-rate, besides a basic dependence upon strain, in the same way as the constant K of eqn (13). The properties of dispersion and of orientation birefringence make it possible to derive relationships that are essential for the evaluation of the optical effect in the plastic range.

The separation of the components of birefringence from, for example, eqn (13) enables the evaluation of the maximum shear stress and of the respective shear strain. In general states, the separation of principal stresses must be effected, by graphical methods of numerical integration, on the basis of the isostatics valid for the loaded state. The separation of the principal normal strains is, however, not a simple matter, since it is first of all necessary to determine the direction of the normal strains. In those cases where the directions of the principal stresses and of the principal strains coincide, separation can be effected with the aid of the evaluation of residual birefringence by means of the measurement of the absolute phase shifts in the principal directions. If the directions of the principal stresses and strains do not coincide, the situation becomes rather complex, and for this case no adequate solution has been devised to date.

Full evaluation can thus proceed along the following lines:

(a) Investigation of the isoclinic pattern under loading and drawing of the trajectories of the principal shear stresses.

(b) Evaluation of the dispersion of birefringence under loading and determination of the maximum shear stress from the equations

$$D = K_D(\gamma - \gamma_p) = k\delta_o = k2K(\gamma - \gamma_P)d$$

$$\delta_o = \frac{D}{k}, \quad \text{where} \quad k = \frac{K_D}{2Kd} \tag{21}$$

$$\delta_s = \delta - \delta_o = 2C\tau d = C(\sigma_1 - \sigma_2)d \tag{22}$$

(The constant K_D (and K) must be substituted with regard to the rate of strain if at different points of the model its value exceeded the permissible tolerance interval.)

(c) Determination of the maximum shear strain:

$$D = K_D(\gamma - \gamma_p)$$

$$\gamma = \frac{D}{K_D} + \gamma_p \tag{23}$$

(d) Determination of the principal normal strains, if the principal stresses and strains coincide at the required points, by means of measurement of the absolute phase shifts in the field of the instantaneous residual birefringence, or separation of the principal stresses by graphical or numerical integration in the case of non-coincidence of principal directions.

The reliability and correctness of measurements of the inelastic states of strain depend in the first place upon precise simulation of the mechanisms of deformation in real materials by model materials, and upon perfect interpretation of the mechanism of deformation by optical phenomena. The linearisation of the relationship between optical effect and the expressions describing the mechanical state is only an approximation, since description of the mechanical state makes use of an accepted theory, which may not in all cases fully reflect the richly interlinked processes. Such an approximation is, nevertheless, acceptable for the solution of technological problems for which other methods— which may have limited adequacy—do not supply such satisfactory results. In solutions of the states of large plastic strains, a theory proceeding from the double nature of birefringence will prove more reliable and more accurate than, or at least as accurate as, any other theory.

Only in order to achieve greater accuracy in the investigation of small elastic-plastic, or anelastic, strains of a comparable order will it be advisable to resort to a theory the evaluation of which is based upon the measurement of different relationships.

8.2. Three-dimensional Investigations

In investigating three-dimensional states the only route is through the application of the Tyndall effect.[28]

The influence of the three-dimensional state of stress upon the birefringent effect and upon the dispersion of birefringence in plastic deformation has not yet been fully investigated in all its aspects. The difficulties are aggravated by the linear polymers being available only in slabs of limited thickness, and the cast crosslinked polymers displaying rather limited plastic behaviour. This is why investigation of these states yields only basic hypotheses proceeding from certain general phenomena.[29–31]

The deformation mechanism of polymers and its relationship to birefringence is not limited in the general case by the three-dimensional state of stress, except in the hydrostatic state of stress, about which no final statement can be made. Nor can reliable judgement be passed on the character of the Tyndall effect for finite strains.

A promising approach is indicated by fixing permanent birefringence in a three-dimensional model by ultra-short wave radiation. Through irradiation by γ-rays at room temperature with a total dose of $2 \cdot 0 \times 10^8$ rads with the aid of a weaker source which enables the irradiation to last for 270 h, birefringence can be fixed in a loaded, imperfectly polymerised, epoxide model. The disadvantage here is that the value of this birefringence decreases logarithmically with time. The method has so far not been applied to linear polymers. In combination with the scattered light method, which would otherwise be difficult to apply because of the force needed for loading, this procedure could find practical application in three-dimensional photoplasticity.

It is obvious that in every three-dimensional problem we must consider the coincidence of the principal stresses and strains. If the principal stresses rotate, the principal directions do not coincide.

8.3. Measuring Procedures

Celluloid, or more generally, materials of the cellulose type, remain the most useful materials for photoplasticity. An outstanding property of these materials is their capacity to scatter light, so that three-dimensional problems can be solved using the Tyndall effect.

The preparation and execution of photoplastic and photoviscoelastic studies require certain specific conditions to be fulfilled, which distinguish this method from the practice of photoelastic investigations. The programme must be based on a profound analysis of the problem under investigation with respect to both the material and the conditions of similarity, and to the general state of stress and strain which is expected to prevail in the elastic-plastic state. Consideration must also be given to the limitations of the method; and a knowledge of these limitations gives assistance in the determination of the points, or parts, of the model upon which measurement and evaluation will concentrate in order to obtain the required results.

In each particular case it is essential to inquire into the influence of creep deformation (which affects the rate of loading) upon the shape of the stress–strain curve, the influence of anisotropy in tension and in compression, and that of coincidence, or non-coincidence, of the principal directions during loading.

The properties of the model material can be modified by a change in the temperature, or the humidity, or by a combination of creep changes and plastic deformation. The different rate of loading influences the values σ_p and $d\sigma/d\varepsilon$ and the influence of time, displaying itself through creep strain, provides a choice of the $\sigma-\varepsilon$ curves. During plastic flow, when the state of stress undergoes a fairly rapid change, the only adequate approach seems to be to determine the isoclinics with the help of a photometric or other photorecording device. The isochromatic lines in monochromatic light must, due to time-dependent changes, also be recorded photographically—a simultaneous check being made upon the points or lines of zero order in white light—or with the aid of the electro-optical method.

Owing to the difficulty of separating the principal normal strains, it is necessary to consider the application of supplementary methods in those cases that require this type of solution. The moiré method is well suited for the purpose: it enables measurement over the whole surface and on the same model on which photoplastic measurement is in progress.

The procedure for the solution of viscoelastic technical problems is described in ref. 33.

The problem remains of how to investigate dynamic time processes. The properties of macromolecular materials have not been investigated to such an extent that the suitability for such measurements could be determined. The occurrence of permanent strain in the vicinity of rapid and of fatigue cracks in these materials, however, is an indication that this particular field of application may yield acceptable results.[26,32]

The interesting mechanical behaviour of some non-transparent polymers in inelastic deformation requires the application of a technique using infrared radiation. Infrared radiation is known to be able to propagate through certain materials that are not transparent to visible radiation. The special method of application consists of adaptation of the source, instrumentation of polarisation, and registration with respect to the model material used. The propagation and absorption of infrared radiation in different materials has been listed elsewhere. The thermal influence and a decrease in optical sensitivity due to an increase in wavelength are the disadvantages of the application of infrared radiation. The necessity of visualising the effect, which requires special, and in most cases expensive, technical equipment, discourages more extensive use of this method in practice.

9. APPLICATION OF METHODS OF INELASTIC POLARISING-OPTICAL EFFECTS

The nature of the photoplasticity of amorphous model materials precludes the modelling of plastic phenomena resulting from dislocation failures, slip in crystal grains, and all changes due to the deformation properties of the polycrystalline structure. What photoplasticity can provide is the determination of the state of elastic-plastic stresses and strains in a plane body under given boundary conditions with due consideration of time effects.

The key problem in all photomechanical methods is the question of the adequacy of its interpretation for mechanical processes. The criteria and aims are the same as in photoelasticity. Assessment of the adequacy of the methods is not normally possible, since both analytical theory and photoviscoelasticity are phenomenological and have got their own weak points.

Of course, photoplasticity, like photoelasticity, cannot be considered the cardinal method of experimental solution for all problems. The advantage of other experimental methods of measuring strains with higher or lower accuracy on the structural material itself is counterbalanced by the fact that instruments capable of measuring both large and small strains lack precision and are limited to measurement at discrete points on the surface only. All these methods, however, are limited by the necessity of interpreting stress from the calibration master curves. They cannot, therefore, cover the effect of constraint. Obtaining a full description of both stresses and strains, which photoplasticity can often provide, ensures its preference over other methods and counterbalances the difficulties of its application.

The most important examples of the application of photoplasticity in engineering deal with concentrations of stress in the vicinity of holes and notches in metallic materials. Hand in hand with the investigation of concentrations at notches goes the investigation of stress fields along cracks and in fractures (see most of the literature in the reference list).

Special applications are concerned with the problems of residual stresses, processes of forming and of extrusion and others.

However, the photoplastic method has its limitations, particularly due to dealing with macromolecular substances when it simulates the plastic deformation of structural materials.

10. INVESTIGATION OF INELASTIC STRAINS USING BIREFRINGENT COATINGS

The principle of the method consists in the photoelastic measurement of birefringence in an applied coating, which interprets strain in the surface of the element under investigation, since it suffers the same deformation as the surface itself.

The application of the method is not limited to elastic deformation or to static loading. Provided the magnitude of the strains to be measured does not lie outside the linear range of the birefringent coating, any non-linear deformations can be subjected to such an investigation procedure. Equipping the polariscope with ultra-rapid recording of the electronic-optical or picture-recording type also makes it possible to investigate dynamic events. If use is made of a coating which retains its properties at high temperatures, investigation of thermal stresses becomes possible.

Special fields of research in inelastic strains are studies of the dynamic and thermal effects. Dynamic problems occur in studies on impacts, rotational states of stress and crack propagation. The problems connected with the viscoelastic character of coatings under the propagation of stress- or strain-waves have not been satisfactorily solved so far. For such cases the evaluation of birefringence has not reached the required level of accuracy. In the investigation of thermal states of stress at high temperatures use can be made of gauges made of natural, transparent, and heat-resistant crystals, such as diamond, sapphire or of refractory glass.

11. PHOTOPLASTICITY OF POLYCRYSTALLINE MODEL MATERIALS

This stimulating and unique method of photoplasticity[1,27] has so far been applied only sporadically and to a few particular cases. The current physical method of investigation of the state of stress of individual crystals is extended to a polycrystal, mostly as a plane body. In order to obtain continuous isochromatics and isoclinics it is necessary to use a sufficiently fine-grained (or textured) structure of cubic crystals (optically isotropic when unloaded) of silver or thallium halide salts. Silver chloride crystals proved to be practically the most suitable. They interpret plastic yield directly and can even model such particular types of behaviour as the Bauschinger effect, quenching, the influence of texturing, the effect of individual grains on the development of a rupture, etc. The optical theory of the method is simple, for birefringence is directly proportional to stress, even in the plastic region; optical sensitivity is good, but a low elasticity modulus limits the method. Texturing, e.g., rolling, introduces anisotropy into the model, but its influence affects the fringe constant only. Evaluation is slightly more complex due to the difference in the parameters of the optical and mechanical isoclinics.

The random distribution of grains and of certain irregularities of the structure are the reason that the isochromatic pattern does not possess such clarity as it does in the case of polymers (Fig. 6). A comparison with isotropic homogeneous materials points to the pronounced influence of the structure. This is due to imperfections in the homogeneity of the sizes of grains, which in some places reach a diameter of 1 mm as compared with regions of homogeneous isochromatics, where their size is about 0·05 mm in specimens of 4 mm thickness (about 40 grains along the light beam is the minimum).

In spite of the attractive features of the method its use has not spread in experimental stress analysis and studies applying it are rare. There may be two causes of this situation. First, it is still rather difficult to produce a polycrystalline material of good quality, viz. fine-grained, homogeneous and possessing the required mechanical parameters.[34]

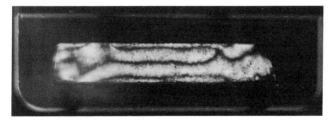

Fig. 6. Isochromatic lines of a four-point loaded beam. Material: fine grained silver chloride, thickness 4 mm. Plastic strains afflict part of the cross-section.

The second cause can be sought in the type of problems best studied by means of this method. Failure mechanics is at present concerned with questions of the state of stress in micro-regions, but photoplasticity is more suited to investigating the influence of the structure of individual grains, or of the thermal conditioning of such phenomena. Micro-scale phenomena, for which the method could also prove very useful, e.g. investigations of quasi-plastic yielding, lie mainly within the confines of the physics of the solid phase. Another field of application could be all kinds of forming and shaping. The application of the method to dynamic processes deserves special attention, particularly with regard to the propagation of plastic strain waves.

REFERENCES

1. Javornický, J., *Photoplasticity*, Elsevier, Amsterdam, 1974.
2. Hayashi, K., Gozo, K. and Yuzo, A., Characterization of mechanical and optical properties of rheo-optically simple materials, *Bull. YSME*, **23,** (180) (1980), 866–73.
3. Doi, M., Molecular rheology of concentrated polymer systems, *J. Polym. Sci., Polym. Phys. Ed.*, **18** (5) (1980), 1005–20.
4. Wenig, W. and Hammel, R., Optical properties and structure of drawn polyethyleneterephthalate–polyethylene films, *Colloid Polymer Sci.*, **260** (1) (1982), 31–6.
5. Vardanyian, G. S., Creep modelling of media with properties changing with time by non-linear similarity method. *Fotouprugost, Sbornik trudov MISI*, (125/126) (1975), 85–90. (In Russian.)
6. Vardanyian, G. S. and Sheremet, V. D., Creep modelling by the method of elastic analogy, *Fotouprugost, Sbornik trudov MISI*, (125/126) (1975), 91–6. (In Russian.)
7. Arnold, W., Call to order by polarising microscope: how to determine the orientation in stretched Plexiglas, *Röhm-Spectrum*, **27** (1983), 69–71.
8. Fujita, K. J., Suehiro, S., Nomura, S. and Kawai, H., Rheo-optical studies on the deformation mechanism of semicrystalline polymers. Uniaxial deformation mechanism of polyethylene spherulites as observed by orientation distribution function of crystallites, *Polymer J.*, **14** (7) (1982), 545–62.
9. Kyu, T., Suehiro, S., Nomura, S. and Kawai, H., Rheo-optical studies on the deformation mechanism of semicrystalline polymers. Orientation retardation spectrum of polyethylene. *J. Polym. Sci., Polym. Phys. Ed.*, **18** (5) (1980), 951–70.
10. Lewis, T. J., Nitrocellulose: birefringence and molecular conformation, *Polymer*, **23** (5) (1982), 710–13.
11. Zachary, I. W. and Riley, W. F., Optical response and yield behaviour of a polyester model material, *Exper. Mech.*, **17** (1977), 321–6.

12. Tougui, A., Gamby, D., Lagarde, A. and Brinson, H. F., Nonlinear photo-viscoelasticity: theory and measurement, *Exper. Mech.*, **23** (1983), 314–21.
13. Antropius, K. and Lind, N. C., A theory of photoelastic effect in glassy-state polymers, *Proc. ICEM 72, Inter. Confer. on Experim. Mechanics, Prague, 1972*, ČVUT Publishing House, Praha, 1974, pp. 10–15.
14. Mönch, E. and Jira, R., Studie zur Photoplastizität von Zelluloid am Rohr unter Innendruck, *Z. angew. Physik*, **7** (9) (1965), 450–3.
15. Mönch, E., Die Dispersion der Doppelbrechung bei Zelluloid als Plastizitätsmass in der Spannungsoptik, *Z. angew. Physik* **6** (8) (1965), 371–5. (In German.)
16. Netrebko, V. P., Dispersion of birefringence in celluloid and its application in photoplasticity, *Vest. Mosk. Univers.*, **3** (1961), 53–60. (In Russian.)
17. Frocht, M. M. and Thompson, R. A., Experiments in mechanical and optical coincidence in photoplasticity, *Proc. SESA*, **18** (1) (1961), 43–7.
18. Ohashi, J. and Nishitani, T., Photorheology, a new method of experimental stress analysis of elastic – viscoelastic body, *Memoirs of the Faculty of Engineering, Nagoya Univ.*, **26** (1) (1974), 53ff.
19. Doyle, J. F., Constitutive relations in photomechanics, *Int. J. Mech. Sci.*, **22** (1980), 1–8.
20. Hunter, A. R., Photoelasto-plastic analysis of notched-bar configuration subjected to bending, *Exper. Mech.*, **10** (1970), 281–7.
21. Zech, J., Ermittlung der Spannungen und Formänderungen bei mechanischen Umformvorgängen an Metallen aus spannungsoptischen Modellversuchen, Doctoral Thesis, Technical University of Clausthal, West Germany, 1976. (In German.)
22. Coleman, B. D. and Dill, E. H., Photoviscoelasticity: theory and practice, In: *Proc. Int. Symp. IUTAM*, ed. J. Kestens, Springer Verlag, Berlin, 1975.
23. Rivlin, R. S. and Smith, G. F., Birefringence in viscoelastic materials, *ZAMP*, **22** (1971) 325–39.
24. Theocaris, P. S., Phenomenological analysis of mechanical and optical behaviour of rheo-optically simple materials, In: *Proc. Int. Symp. IUTAM*, ed. J. Kestens, Springer-Verlag, Berlin, 1975, pp. 146–230.
25. Sharafutdinov, G. Z., Use of nonlinear viscoelastic materials in the polarization-optical method, *Mech. Solids*, **17** (4) (1982), 138–44.
26. Read, B. E., Viscoelastic behaviour of amorphous polymers in the glass–rubber transition region: birefringence studies, *Polymers Eng. Sci.*, **23** (1983), 835–43.
27. Javornický, J., On photoplasticity and its contemporary prospect, *Proc. IUTAM Symp. Optical Methods in Mechanics of Solids, Poitiers, 1973*, ed. A. Lagarde, Sijthoff and Noordhoff, Amsterdam, 1981, pp. 587–611.
28. Johnson, R. L., Measurement of elastic-plastic stresses by scattered-light photomechanics, *Exper. Mech.*, **16** (1976), 201–8.
29. de Franca Freire, J. L., Application of three-dimensional photoplasticity to plane strain and axisymmetric compression problems, *Proc. 4th Inter. Congr. Exper. Mech.*, Boston, 1980.
30. Gomide, H. A. and Burger, C. P., Three-dimensional strain distribution in

upset rings by photoplastic simulation, *Proc. 4th Inter. Congr. Exper. Mech.*, Boston, 1980.

31. Chen, P. C. T. and Cheng, Y. F., Stress analysis of an overloaded breech ring, *Proc. Int. Conf. on Reliab. Stress Anal. and Failure Prev.*, San Francisco, *1980*, ASME, New York, pp. 175–80.
32. Mönch, E., Betz, A. and Kuch, R., Zur Frage der dynamischen Photoplastizität, *Proc. 7th Inter. Conf. on Exper. Stress Analysis*, Haifa, 1982.
33. Laermann, K. H., Problems in application of photoviscoelasticity in practical experimental stress analysis, *The Von Karmán Session, Internat. Centre for Mechanical Sciences*, Udine, Italy, 1983.
34. Dietz, P., Hirchenhain, A. and Schmidt, O. A., Spannungsoptische Modelluntersuchungen elasto-plastischer und plastischer Vorgänge mit Hilfe des Materials Silberchlorid. Verfahren, Vergleiche, Anwendungen, 6. *GESA—Symposium 'Experimentelle Spannungsanalyse'*, Fellbach b. Stuttgart, 1982.

3

INTEGRATED PHOTOELASTICITY OF AXISYMMETRIC CUBIC SINGLE CRYSTALS

H. K. Aben

Institute of Cybernetics, Academy of Sciences of the Estonian SSR, Tallinn, USSR

ABSTRACT

In this chapter, the application of integrated photoelasticity as a non-destructive means of determining the residual stresses in axisymmetric cubic single crystals is described. The single crystal is put into an immersion bath and polarised light is passed through it at a plane perpendicular to the crystal axis, parallel to a crystallographic axis and at an angle of 45° to it. For a number of rays, the characteristic directions as well as the characteristic phase retardation are determined. The stress components are presented as power series of the dimensionless radial coordinate. Algorithms have been developed, permitting the determination of the coefficients of these expansions on the basis of the experimental data. For the separation of the principal stresses the equation of equilibrium is applied. As examples, stress determination in an NaCl cylinder without axial stress gradient, and in a KCl cylinder with axial stress gradient, are described.

1. INTRODUCTION

Single crystals are widely used as elements of infrared optics, laser crystals, windows of high-power lasers, scintillators, acoustic crystals, light modulators of large aperture, etc. Their quality as optical elements depends to a great extent on the internal stresses which arise during the growing process. A non-destructive method which permits determination of the residual stresses is indispensable in order to elaborate a suitable growing technology and to check the quality of single crystals.

The general theory of photoelasticity of crystals is presented in refs 1–4. Determination of stresses in crystal plates cut perpendicular to a crystallographic axis is considered in refs 5–11. In refs 12–14 the methods of determining stresses in plates cut out arbitrarily from a cubic crystal are considered. Several authors[15–21] have used photoelasticity to determine stresses around dislocations. A review of new applications and generalisations of the photoelasticity of crystals is given in ref. 22.

Methods of calculating technological quenching stresses in crystals of cylindrical form have been developed by several authors.[23–28] However, in practice it is hard to obtain reliable values for the stresses, because one has no exact data about either the temperature field or the change of the elastic and physical properties of the crystal during the growing process. Therefore experimental methods should be used for this purpose.

Chernysheva[29] has proposed a simplified method of determining stresses in cubic single crystals of cylindrical form on the basis of integral phase retardation, measured when light passes the cylinder perpendicular to its axis. The method is based on the assumption that the axial stress is distributed according to a simple parabola. This method may yield erroneous results since the stress distribution in single crystals depends strongly on the parameters of the growing process and may differ considerably from a simple parabola.[30]

Hornstra and Penning,[31] and Nikitenko and Indenbom[32] have proposed the following method for experimental determination of stress in cubic single crystals of cylindrical form: perpendicular to the cylinder axis a plate is cut out of the crystal and its stresses are determined by the aid of two-dimensional photoelasticity. With certain assumptions the stresses in the whole cylinder can now be calculated. For the same purpose, the plate to be investigated may be cut out of the cylinder in the longitudinal direction.

The drawbacks of this method are, firstly, that it is destructive, and, secondly, that due to rather critical additional assumptions it may give inaccurate results.

There are only a few papers[33–35] devoted to the application of the scattered light method for stress determination in single crystals. One of the difficulties connected with this is that in many crystals the stress-birefringence is very low.

The simplest method of determining three-dimensional stress distribution is integrated photoelasticity.[36] In this case polarised light is passed through the whole model (usually while the model is in an immersion bath) and the stresses are determined on the basis of the integral optical phenomenon (integral fringe pattern, characteristic directions, etc.). Integrated photoelasticity can be applied when the model has certain properties of symmetry. Single crystals are often grown in the form of cylinders, or prisms with quadratic cross-section. In the former case the stress distribution is practically axisymmetric; in the latter case, the stress distribution has four planes of symmetry. Therefore, single crystals are often suitable objects for integrated photoelasticity.

During recent years integrated photoelastic methods for determining stresses in cubic single crystals of cylindrical and prismatic form have been developed[37-44] and successfully applied to checking the quality of single crystals and developing suitable growing technology for them. The aim of this chapter is to give a systematic and detailed description of the method of determining stresses in cubic single crystals of axisymmetric form.

2. BASIC FORMULAE

Let the components of the electric vector E_i and that of the electric induction D_j be connected by the relation

$$E_i = a_{ij}D_j \tag{1}$$

where a_{ij} are components of the index tensor[4] with respect to the coordinate axes chosen. The equation of the index ellipsoid is

$$a_{11}x^2 + a_{22}y^2 + a_{33}z^2 + 2a_{23}yz + 2a_{31}zx + 2a_{12}xy = 1 \tag{2}$$

or in principal axes

$$a_1x^2 + a_2y^2 + a_3z^2 = 1 \tag{3}$$

where

$$a_i = \frac{1}{n_i^2} = \frac{1}{\varepsilon_i} \tag{4}$$

Here n_i are the principal refractive indices, and ε_i are the principal values of the dielectric tensor.

We shall denote the value of a_{ij} in the undeformed crystal by a_{ij}^0 and if it changes to a_{ij} on deformation, then the change

$$\Delta a_{ij} = a_{ij} - a_{ij}^0 \tag{5}$$

can be expressed as a homogeneous linear function of the stress components

$$\Delta a_i = \sum \pi_{ij}\sigma_j \tag{6}$$

where π_{ij} are the piezo-optic constants and σ_j are stress components in matrix notation according to

$$\begin{pmatrix} \sigma_{11} & \sigma_{12} & \sigma_{13} \\ \sigma_{21} & \sigma_{22} & \sigma_{23} \\ \sigma_{31} & \sigma_{32} & \sigma_{33} \end{pmatrix} \rightarrow \begin{pmatrix} \sigma_1 & \sigma_6 & \sigma_5 \\ \sigma_6 & \sigma_2 & \sigma_4 \\ \sigma_5 & \sigma_4 & \sigma_3 \end{pmatrix} \tag{7}$$

The constants π_{ij} are sometimes referred to as photoelastic constants or stress-optic constants.

In the following we shall be considering cubic single crystals of classes $\bar{4}3m$, 432, and $m3m$, in which case the tensor π_{ij} is

$$\begin{pmatrix} \pi_{11} & \pi_{12} & \pi_{12} & 0 & 0 & 0 \\ \pi_{12} & \pi_{11} & \pi_{12} & 0 & 0 & 0 \\ \pi_{12} & \pi_{12} & \pi_{11} & 0 & 0 & 0 \\ 0 & 0 & 0 & \pi_{44} & 0 & 0 \\ 0 & 0 & 0 & 0 & \pi_{44} & 0 \\ 0 & 0 & 0 & 0 & 0 & \pi_{44} \end{pmatrix} \tag{8}$$

In integrated photoelasticity of single crystals it is expedient to express the stress tensor σ_{ij} in the crystallographic axes, but the index tensor a'_{ij} in axes determined by the direction of light propagation in the crystal. The prime denotes that the tensors σ_{ij} and a'_{ij} are expressed in different coordinate systems.

Let us assume that the light passes through a layer of a single crystal in the plane (001) with the wave normal to the x'_2 direction (Fig. 1). The polarisation of light is influenced by the components of the index tensor a'_{11}, a'_{33} and a'_{13}. The latter can be expressed as

$$\Delta a'_{11} = [\pi_{11}\sigma_{11} + \pi_{12}(\sigma_{22} + \sigma_{33})]\cos^2\beta$$
$$+ [\pi_{12}(\sigma_{11} + \sigma_{33}) + \pi_{11}\sigma_{22}]\sin^2\beta + \pi_{44}\sigma_{12}\sin 2\beta \tag{9}$$

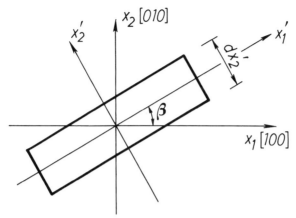

Fig. 1. The stress tensor is expressed in the crystallographic axes, x_1, x_2 and x_3, the index tensor a'_{ij} in the axes x'_1, x'_2, x'_3 determined by the direction x'_2 of the wave normal.

$$\Delta a'_{33} = \pi_{12}(\sigma_{11} + \sigma_{22}) + \pi_{11}\sigma_{33} \tag{10}$$

$$\Delta a'_{13} = \pi_{44}(\sigma_{13}\cos\beta + \sigma_{23}\sin\beta) \tag{11}$$

For the components of the index tensor in the plane $x'_1 x'_3$ the following relations hold

$$a'_1 - a'_2 = \frac{a'_{11} - a'_{22}}{\cos 2\varphi'} = \frac{2a'_{13}}{\sin 2\varphi'} \tag{12}$$

where a'_1 and a'_2 are the principal values of the index tensor in the plane $x'_1 x'_3$, and φ' is the azimuth of the principal direction.

Optical path difference is determined through the difference of refractive indices. Since usually

$$\Delta a_{ij} \ll 1 \tag{13}$$

we have

$$\Delta n' = \tfrac{1}{2}n_0^3(a'_1 - a'_2) \tag{14}$$

where n_0 is the refractive index of the nonstressed crystal.

According to eqns (9)–(14) the optical path difference δ' and the parameter of the isoclinic φ' are related to the stresses by the following expressions[39]

$$\delta'\cos 2\varphi' = f_1\,dx'_2 \tag{15}$$

$$\delta'\sin 2\varphi' = f_2\,dx'_2 \tag{16}$$

where

$$f_1 = \tfrac{1}{2}n_0^3[(\pi_{11} - \pi_{12})(\sigma_{11}\cos^2\beta + \sigma_{22}\sin^2\beta - \sigma_{33})$$
$$+ \pi_{44}\sigma_{12}\sin 2\beta] \tag{17}$$

$$f_2 = n_0^3\pi_{44}(\sigma_{31}\cos\beta + \sigma_{23}\sin\beta) \tag{18}$$

and dx_2' is the thickness of the layer.

Another, the more traditional form of eqns (17) and (18), is

$$\delta' = \sqrt{f_1^2 + f_2^2}\, dx_2' \tag{19}$$

$$\tan 2\varphi' = \frac{f_2}{f_1} \tag{20}$$

Equation (20) shows that in a crystal the parameters of the mechanical and optical isoclinics may be different.

Assume that the axis of a crystal of cylindrical form is parallel to the crystallographic axis $x_3 = x_3'$ and there is no stress gradient in that direction. This is the case in the middle part of comparatively long crystals. Then we have $\sigma_{31} = \sigma_{23} = 0$. Equations (18) and (20) yield that now $\cos 2\varphi' = 1$ and $\sin 2\varphi' = 0$.

Since there is no rotation of the principal axes on a light ray, the integral Wertheim law[36] may be used and, taking into account eqns (15) and (17), the integral optical path difference is

$$\delta' = \tfrac{1}{2}n_0^3\int_{x_{20}'}^{x_{2*}'} [(\pi_{11} - \pi_{12})(\sigma_{11}\cos^2\beta + \sigma_{22}\sin^2\beta - \sigma_{33})$$
$$+ \pi_{44}\sigma_{12}\sin 2\beta]\, dx_2' \tag{21}$$

where x_{20}' and x_{2*}' are the coordinates of the points where the light ray enters the model and emerges from it.

Equation (21) is the basic formula in integrated photoelasticity for cubic single crystals, when the axial stress gradient can be neglected.

If a stress gradient in the direction of the crystal axis x_3 is present, the stresses σ_{31} and σ_{23} are not equal to zero. This causes, in general, rotation of the secondary principal axes on the light ray and the integral Wertheim law is not valid. According to Brosman[38] in this case the photoelastic phenomena can be treated approximately, as follows.

Let us represent the single crystal as a pile of m birefringent plates.

Each plate j can be described by a matrix[45] U_j

$$U_j = \cos \frac{\gamma_j}{2} \left[\begin{pmatrix} 1 & 0 \\ 0 & 1 \end{pmatrix} + i \begin{pmatrix} \cos 2\varphi_j & \sin 2\varphi_j \\ \sin 2\varphi_j & -\cos 2\varphi_j \end{pmatrix} \tan \frac{\gamma_j}{2} \right] \qquad (22)$$

where γ_j is the phase retardation and φ_j is the azimuth of the fast axis. The matrix U of the whole single crystal between the points of entrance and emergence of light is

$$U = \prod_{j=1}^{m} U_j \qquad (23)$$

If the thickness of each plate is small, we have

$$\tan \frac{\gamma_j}{2} \cong \frac{\gamma_j}{2}, \quad \cos \frac{\gamma_j}{2} = 1 \qquad (24)$$

Let us introduce eqn (22) into eqn (23) and neglect higher order terms relative to phase retardations. In the limiting case $m \to \infty$ we obtain

$$U = \begin{pmatrix} 1 + iq & is \\ is & 1 - iq \end{pmatrix} \qquad (25)$$

where

$$q = \frac{180}{\lambda} \int_{x'_{20}}^{x'_{2*}} f_i \, dx'_2 \qquad (26)$$

$$s = \frac{180}{\lambda} \int_{x'_{20}}^{x'_{2*}} f_2 \, dx'_2 \qquad (27)$$

Since the matrix (eqn (25)) is symmetric, it represents a birefringent plate with optical path difference δ' and an azimuth of the principal directions φ'. Now the following relations hold

$$\delta' \cos 2\varphi' = \frac{\lambda}{180} q = \int_{x'_{20}}^{x'_{2*}} f_1 \, dx'_2 \qquad (28)$$

$$\delta' \sin 2\varphi = \frac{\lambda}{180} s = \int_{x'_{20}}^{x'_{2*}} f_2 \, dx'_2 \qquad (29)$$

The approximate eqns (28) and (29) were derived by Brosman[38] for the case of cubic single crystals on the basis of the paper by Tronko and Golovach.[45] At about the same time a similar result was published for the case of the integrated photoelasticity of isotropic axisymmetric bodies by Doyle and Danyluk.[46]

Equations (28) and (29) are basic formulae in integrated photoelasticity for axisymmetric cubic single crystals with axial stress gradient when either the birefringence is low or the rotation of the secondary principal axes on the light ray is small.

3. CRYSTALS OF CYLINDRICAL FORM WITHOUT AXIAL STRESS GRADIENT

Let us consider the determination of stress in a crystal cylinder with no axial stress gradient. It is assumed that the stress distribution is axisymmetric.

Both assumptions are justified by determining residual quenching stresses which are of chief interest in the photoelasticity of single crystals. Experimental investigations have shown that in single crystals of cylindrical form, which are grown in stationary conditions, the axial gradient of the residual stresses is low, except for the end regions of the crystal. As an example, in Fig. 2, the integral fringe pattern of an NaCl single crystal of cylindrical form, grown from the melt, is shown.

During the growing process the temperature field is usually axisymmetric. Grechushnikov and Brodovskii,[23] as well as Sirotin[24] have shown that in this case residual thermal stresses in cubic crystals of cylindrical form are also axisymmetric. Often cubic single crystals are optically strongly anisotropic, while their mechanical anisotropy is weak (e.g. garnets[47]). Therefore, with axisymmetric loads, one may expect that the stress distribution is also approximately axisymmetric.

If light is passed through a cross-section of the crystal parallel to an arbitrary direction x_2' (Fig. 3), the secondary principal directions are parallel and perpendicular to the crystal axis and the integral optical path difference δ' is expressed by eqn (21).

From the equilibrium condition of a layer of the segment of the cylinder in the x_1' direction it follows that

$$\int_{x_{20}'}^{x_{2*}'} (\sigma_{11} \cos^2 \beta + \sigma_{22} \sin^2 \beta + \sigma_{12} \sin 2\beta) \, dx_2' = 0 \qquad (30)$$

We should mention that an analogous equilibrium condition was

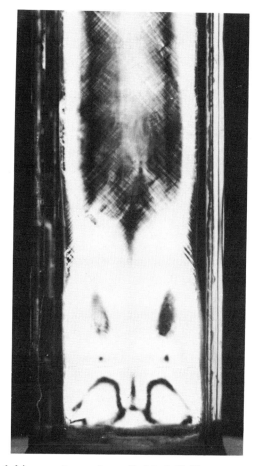

Fig. 2. Integral fringe pattern of a cylindrical NaCl crystal grown from the melt.

used in the integrated photoelasticity of isotropic cylinders by Poritsky.[49]

Putting eqn (30) into eqn (21) yields

$$\delta' = \tfrac{1}{2} n_0^3 \int_{x'_{20}}^{x'_{2*}} \{[\pi_{44} - (\pi_{11} - \pi_{12})]\sigma_{12} \sin 2\beta$$
$$- (\pi_{11} - \pi_{12})\sigma_{33}\} \, dx'_2 \tag{31}$$

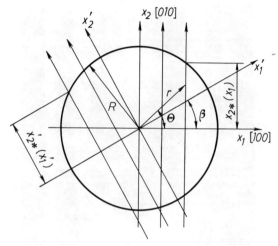

Fig. 3. Passing of light through a cross-section of a cylindrical crystal.

Taking into account the following relations

$$\sigma_{11} = \sigma_r \cos^2 \theta + \sigma_\theta \sin^2 \theta$$

$$\sigma_{22} = \sigma_\theta \cos^2 \theta + \sigma_r \sin^2 \theta$$

$$\sigma_{33} = \sigma_z \tag{32}$$

$$\sigma_{12} = (\sigma_r - \sigma_\theta) \sin \theta \cos \theta$$

eqn (31) in cylindrical coordinates is

$$\delta' = \tfrac{1}{2} n_0^3 \int_{x'_{20}}^{x'_{2^*}} \{[\pi_{44} - (\pi_{11} - \pi_{12})](\sigma_r - \sigma_\theta) \sin \theta \cos \theta \sin 2\beta$$

$$- (\pi_{11} - \pi_{12}) \sigma_z \} \, dx'_2 \tag{33}$$

If light passes the cross-section parallel to the crystallographic axis x_2, i.e. $\beta = 0$, eqn (33) yields

$$\delta(x_1) = -n_0^3 (\pi_{11} - \pi_{12}) \int_0^{x_{2^*}} \sigma_z \, dx_2 \tag{34}$$

The latter equation may be written as[48]

$$\delta(x_1) = -n_0^3 (\pi_{11} - \pi_{12}) \int_{x_1}^{R} \sigma_z \frac{r \, dr}{\sqrt{r^2 - x_1^2}} \tag{35}$$

Thus, in this case the optical path difference is determined only by the axial stress.

Equation (35) is the Abel integral equation and its solution is

$$\sigma_z(r) = -\frac{2}{\pi r n_0^3(\pi_{11} - \pi_{12})} \frac{d}{dr} \int_R^r \frac{x_1\,\delta(x_1)\,dx_1}{\sqrt{r^2 - x_1^2}} \tag{36}$$

An analogous problem arises in the interferometry of axisymmetric phase objects.[50] In the latter case the highest disturbance usually takes place near the axis of the phase object and the Abel inversion gives satisfactory results. In cylindrical single crystals maximum stresses usually occur near the boundary. In this case eqn (36) cannot be used effectively and we consider another approach.

Let us express σ_z as

$$\sigma_z = \sum_{k=0}^{l} b_{2k}\rho^{2k} \tag{37}$$

where b_{2k} are coefficients to be determined and ρ is a dimensionless radial coordinate

$$\rho = \frac{r}{R} \tag{38}$$

Equation (35) may be expressed as

$$\delta(\xi) = -n_0^3(\pi_{11} - \pi_{12})R \int_\xi^1 \sigma_z \frac{\rho\,d\rho}{\sqrt{\rho^2 - \xi^2}} \tag{39}$$

where

$$\xi = \frac{x_1}{R} \tag{40}$$

If we introduce eqn (37) into eqn (39) and integrate, we obtain

$$\frac{\delta(\xi)}{n_0^3(\pi_{11} - \pi_{12})R} = \sum_{k=0}^{l} b_{2k}G_{2k}(\xi) \tag{41}$$

where

$$G_0(\xi) = \sqrt{1 - \xi^2}$$

$$G_2(\xi) = \tfrac{1}{3}\sqrt{1 - \xi^2}(1 + 2\xi^2)$$

$$G_4(\xi) = \tfrac{1}{5}\sqrt{1 - \xi^2}(1 + \tfrac{4}{3}\xi^2 + \tfrac{8}{3}\xi^4) \tag{42}$$

$$G_6(\xi) = \tfrac{1}{7}\sqrt{1 - \xi^2}(1 + \tfrac{6}{5}\xi^2 + \tfrac{8}{5}\xi^4 + \tfrac{16}{5}\xi^6)$$

and the following functions $G_{2k}(\xi)$ can be calculated recursively by the aid of the formula

$$G_{2k}(\xi) = \frac{\sqrt{1-\xi^2}}{(2k+1)} + \frac{2k}{2k+1} \xi^2 G_{2k-2}(\xi) \tag{43}$$

Values of the functions $G_0(\xi)$, $G_2(\xi)$, $G_4(\xi)$ and $G_6(\xi)$ are given in Table 1 and graphically shown in Fig. 4. These functions may be interpreted as distributions of the integral optical path difference due to separate terms in the eqn (37).

To determine the coefficients b_{2k} in eqn (37), the integral optical path difference should be measured for m ($m \geq l$) values of ξ. Equation (41) now yields m equations

$$\frac{\delta(\xi_i)}{n_0^3(\pi_{11} - \pi_{12})R} = \sum_{k=0}^{l} b_{2k} G_{2k}(\xi_i) \qquad i = 1, 2, \ldots, m \tag{44}$$

TABLE 1
Functions $G_{2k}(\xi)$

ξ	G_0	G_2	G_4	G_6
0·00	1·0000	0·3333	0·2000	0·1428
0·05	0·9987	0·3345	0·2004	0·1431
0·10	0·9949	0·3382	0·2017	0·1438
0·15	0·9886	0·3443	0·2039	0·1451
0·20	0·9797	0·3527	0·2072	0·1470
0·25	0·9682	0·3630	0·2118	0·1496
0·30	0·9539	0·3752	0·2178	0·1530
0·35	0·9367	0·3887	0·2254	0·1574
0·40	0·9165	0·4032	0·2349	0·1631
0·45	0·8930	0·4182	0·2463	0·1703
0·50	0·8660	0·4330	0·2598	0·1793
0·55	0·8351	0·4468	0·2751	0·1906
0·60	0·8000	0·4586	0·2920	0·2044
0·65	0·7599	0·4673	0·3099	0·2208
0·70	0·7141	0·4713	0·3275	0·2396
0·75	0·6614	0·4685	0·3431	0·2599
0·80	0·6000	0·4560	0·3534	0·2796
0·85	0·5267	0·4293	0·3535	0·2941
0·90	0·4358	0·3806	0·3338	0·2940
0·95	0·3122	0·2919	0·2732	0·2559
1·00	0·0000	0·0000	0·0000	0·0000

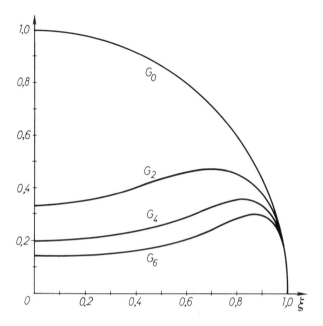

Fig. 4. Functions $G_0(\xi)$, $G_2(\xi)$, $G_4(\xi)$ and $G_6(\xi)$.

from which the coefficients b_{2k} can be determined with the aid of the least squares method.

Thus, the axial stress distribution has been determined. Application of the so-called sum rule

$$p\sigma_z = \sigma_r + \sigma_\theta \tag{45}$$

and integration of the equilibrium equation permits the determination of all the stresses analogously to the isotropic case.[48] However, depending on the crystal growing technology and on heat treatment, the parameter p in eqn (45) may be equal to 1, -1, μ or $1/\mu$ (μ is Poisson's ratio).[21,51] Therefore, the exact value of p is not usually known and it is better to avoid application of the sum rule.

In order to determine all the stresses without application of the sum rule we need additional experimental data. For this let us pass light through the cross-section of the crystal parallel to the x_2' axis with

$\beta = 45°$ (Fig. 3). In this case eqn (4) yields

$$\delta'(x_1') = \tfrac{1}{2}n_0^3 \int_{x_{20}'}^{x_{2*}'} \{[\pi_{44} - (\pi_{11} - \pi_{12})](\sigma_r - \sigma_\theta)\sin\theta\cos\theta$$
$$- (\pi_{11} - \pi_{12})\sigma_z\}\,dx_2' \tag{46}$$

From eqns (46) and (34) follows

$$\frac{[\delta'(\xi') - \delta(\xi)]_{\xi'=\xi}}{n_0^3[\pi_{44} - (\pi_{11} - \pi_{12})]R} = \int_0^{\sqrt{1-\xi^2}} (\sigma_r - \sigma_\theta)\sin\theta\cos\theta\,d\eta' \tag{47}$$

where

$$\xi' = \frac{x_1'}{R}, \qquad \eta' = \frac{x_2'}{R} \tag{48}$$

The latter equation shows that the difference between the optical path differences, measured when light passes through the model, once parallel to the x_2 axis, and then parallel to the x_2' axis (Fig. 3, $\beta = 45°$), at the same distance $\xi' = \xi$ from the cylinder axis, is determined through the difference in the radial and tangential stresses $\sigma_r - \sigma_\theta$.

It is important to notice that eqn (47) can only be used in the case of single crystals. In the case of an isotropic cylinder $\delta'(\xi) = \delta(\xi)$, the passing of light through the cylinder in several directions does not give any new information about the stresses. Thus, although the integrated photoelasticity of single crystals is more complicated than that of isotropic bodies, it also opens up some new possibilities for determining stresses.

Let us express $\sigma_r - \sigma_\theta$ as

$$\sigma_r - \sigma_\theta = \sum_{k=1}^{l} c_{2k}\rho^{2k} \tag{49}$$

If we introduce eqn (49) into eqn (47) and integrate, we obtain

$$\frac{\delta'(\xi) - \delta(\xi)}{n_0^3[\pi_{44} - (\pi_{11} - \pi_{12})]R} = \sum_{k=1}^{m} c_{2k}F_{2k}(\xi) \tag{50}$$

where

$$F_2(\xi) = \tfrac{1}{3}\sqrt{1-\xi^2}(-1+4\xi^2)$$

$$F_4(\xi) = \tfrac{1}{5}\sqrt{1-\xi^2}(-1+2\xi^2+4\xi^4) \tag{51}$$

$$F_6(\xi) = \tfrac{1}{7}\sqrt{1-\xi^2}(-1+\tfrac{8}{5}\xi^2+\tfrac{32}{15}\xi^4+\tfrac{64}{15}\xi^6)$$

TABLE 2
Functions $F_{2k}(\xi)$

ξ	F_2	F_4	F_6
0·00	−0·3333	−0·2000	−0·1428
0·05	−0·3295	−0·1987	−0·1421
0·10	−0·3183	−0·1949	−0·1398
0·15	−0·2999	−0·1884	−0·1359
0·20	−0·2743	−0·1790	−0·1304
0·25	−0·2420	−0·1664	−0·1231
0·30	−0·2035	−0·1502	−0·1138
0·35	−0·1592	−0·1302	−0·1022
0·40	−0·1099	−0·1058	−0·0879
0·45	−0·0565	−0·0769	−0·0705
0·50	0·0000	−0·0433	−0·0494
0·55	0·0584	−0·0048	−0·0241
0·60	0·1173	0·0381	0·0058
0·65	0·1747	0·0849	0·0411
0·70	0·2285	0·1343	0·0814
0·75	0·2755	0·1839	0·1260
0·80	0·3120	0·2302	0·1728
0·85	0·3318	0·2668	0·2166
0·90	0·3254	0·2828	0·2467
0·95	0·2716	0·2537	0·2372
1·00	0·0000	0·0000	0·0000

and the following functions $F_{2k}(\xi)$ can be calculated recursively from

$$F_{2k}(\xi) = \frac{2k+2}{2k+1}\xi^2 G_{2k-2}(\xi) - \frac{\sqrt{1-\xi^2}}{2k+1} \qquad (52)$$

Values of the functions $F_2(\xi)$, $F_4(\xi)$, and $F_6(\xi)$ are given in Table 2 and are shown graphically in Fig. 5.

If integral optical path differences $\delta(\xi)$ and $\delta'(\xi)$ have been measured for n ($n \geqslant m$) values of ξ, eqn (50) yields a system of n equations

$$\frac{\delta'(\xi_i) - \delta(\xi_i)}{n_0^3[\pi_{44} - (\pi_{11} - \pi_{12})]R} = \sum_{k=1}^{m} c_{2k}F_{2k}(\xi_i) \qquad i = 1, 2, \ldots, n \qquad (53)$$

from which the coefficients c_{2k} can be calculated with the aid of the least squares method.

In order to determine the radial stress we have to integrate the

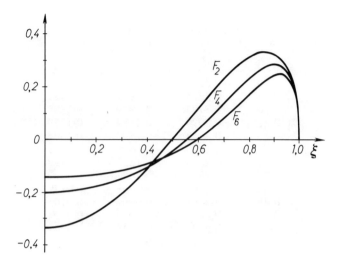

Fig. 5. Functions $F_2(\xi)$, $F_4(\xi)$ and $F_6(\xi)$.

equilibrium equation. We obtain

$$\sigma_r = -\int_1^\rho \frac{\sigma_r - \sigma_\theta}{\rho} \, d\rho \qquad (54)$$

Substituting eqn (49) into eqn (54) and integrating yields

$$\sigma_r = \tfrac{1}{2}c_2(1-\rho^2) + \tfrac{1}{4}c_4(1-\rho^4) + \tfrac{1}{6}c_6(1-\rho^6) + \cdots \qquad (55)$$

Introducing eqn (55) into eqn (49) gives

$$\sigma_\theta = \tfrac{1}{2}c_2(1-3\rho^2) + \tfrac{1}{4}c_4(1-5\rho^4) + \tfrac{1}{6}c_6(1-7\rho^6) + \cdots \qquad (56)$$

Now all the stress components have been determined and the solution in the general form is

$$\sigma_z = \sum_{k=0}^{l} b_{2k}\rho^{2k} \qquad (57)$$

$$\sigma_r = \sum_{k=1}^{l+1} \frac{1}{2k} c_{2k}(1-\rho^{2k}) \qquad (58)$$

$$\sigma_\theta = \sum_{k=1}^{l+1} \frac{1}{2k} c_{2k}[1-(2k+1)\rho^{2k}] \qquad (59)$$

4. THE GENERAL CASE

If an axial stress gradient is present, we have to use eqns (28) and (29), where f_1 and f_2 are determined by eqns (17) and (18). When $\beta = 0$, eqn (18) yields

$$f_2 = n_0^3 \pi_{44} \sigma_{31} \tag{60}$$

or in cylindrical coordinates

$$f_2 = n_0^3 \pi_{44} \sigma_{rz} \cos \theta \tag{61}$$

Equation (29) now becomes

$$\delta \sin 2\varphi = n_0^3 \pi_{44} \int_{x_{20}}^{x_{2*}} \sigma_{rz} \cos \theta \, dx_2 \tag{62}$$

Let us express σ_{rz} as

$$\sigma_{rz} = \sum_{k=1}^{l} d_{2k+1}\rho(1 - \rho^{2k}) \tag{63}$$

This function (eqn (63)) satisfies the boundary conditions since $\sigma_{rz} = 0$ at the centre of the cross-section as well as at the boundary. Inserting eqn (63) into eqn (62) yields

$$\delta \sin 2\varphi = 2n_0^3 \pi_{44} R \sum_{k=1}^{l} d_{2k+1} \int_0^{\sqrt{1-\xi^2}} \rho(1 - \rho^{2k}) \cos \theta \, d\eta \tag{64}$$

where

$$\eta = \frac{x_2}{R}$$

Integrating eqn (64) gives

$$\frac{\delta(\xi) \sin 2\varphi(\xi)}{2n_0^3 \pi_{44} R} = \sum_{k=1}^{l} d_{2k+1} I_k(\xi) \tag{65}$$

where

$$I_1(\xi) = \tfrac{2}{3}\xi\sqrt{1 - \xi^2}(1 - \xi^2)$$

$$I_2(\xi) = \tfrac{4}{5}\xi\sqrt{1 - \xi^2}(1 - \tfrac{1}{3}\xi^2 - \tfrac{2}{3}\xi^4)$$

$$I_3(\xi) = \tfrac{6}{7}\xi\sqrt{1 - \xi^2}(1 - \tfrac{1}{5}\xi^2 - \tfrac{4}{15}\xi^4 - \tfrac{8}{15}\xi^6)$$

$$I_4(\xi) = \tfrac{8}{9}\xi\sqrt{1 - \xi^2}(1 - \tfrac{1}{7}\xi^2 - \tfrac{6}{35}\xi^4 - \tfrac{8}{35}\xi^6 - \tfrac{16}{35}\xi^8)$$

$$\tag{66}$$

and in general

$$I_k(\xi) = \xi\sqrt{1-\xi^2} - I_k'(\xi) \tag{67}$$

$$I_k'(\xi) = \frac{\xi\sqrt{1-\xi^2}}{2k+1} + \frac{2k}{2k+1}\,\xi^2 I_{k-1}'(\xi) \tag{68}$$

$$I_1'(\xi) = \tfrac{1}{3}\xi\sqrt{1-\xi^2}(1+2\xi^2) \tag{69}$$

Values of the functions $I_1(\xi)$, $I_2(\xi)$, $I_3(\xi)$ and $I_4(\xi)$ are given in Table 3 and are shown graphically in Fig. 6.

Coefficients d_{2k+1} of eqn (63) are determined from the system of equations

$$\frac{\delta(\xi_i)\sin 2\varphi(\xi_i)}{2n_0^3\pi_{44}R} = \sum_{k=1}^{l} d_{2k+1}I_k(\xi_i) \qquad i = 1, 2, \ldots, m \quad (m \geqslant l) \tag{70}$$

by the method of least squares.

TABLE 3
Functions $I_{2k}(\xi)$

ξ	I_1	I_2	I_3	I_4
0·00	0·0000	0·0000	0·0000	0·0000
0·05	0·0332	0·0399	0·0427	0·0444
0·10	0·0657	0·0793	0·0851	0·0883
0·15	0·0966	0·1177	0·1265	0·1314
0·20	0·1254	0·1545	0·1665	0·1731
0·25	0·1513	0·1891	0·2046	0·2131
0·30	0·1736	0·2208	0·2403	0·2507
0·35	0·1918	0·2489	0·2727	0·2854
0·40	0·2053	0·2726	0·3013	0·3166
0·45	0·2137	0·2910	0·3252	0·3434
0·50	0·2165	0·3031	0·3433	0·3650
0·55	0·2136	0·3080	0·3545	0·3801
0·60	0·2048	0·3047	0·3573	0·3874
0·65	0·1902	0·2925	0·3504	0·3852
0·70	0·1670	0·2706	0·3321	0·3713
0·75	0·1447	0·2387	0·3011	0·3435
0·80	0·1152	0·1972	0·2563	0·2994
0·85	0·0828	0·1473	0·1977	0·2374
0·90	0·0497	0·0918	0·1276	0·1582
0·95	0·0193	0·0371	0·0535	0·0686
1·00	0·0000	0·0000	0·0000	0·0000

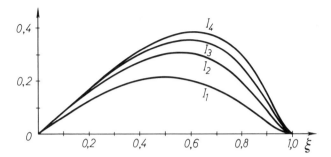

Fig. 6. Functions $I_1(\xi)$, $I_2(\xi)$, $I_3(\xi)$ and $I_4(\xi)$.

Now let us consider determination of $\sigma_r - \sigma_\theta$. If $\beta = 0$, eqn (15) yields

$$\delta \cos 2\varphi = \tfrac{1}{2}n_0^3(\pi_{11} - \pi_{12}) \int_{x_{20}}^{x_{2*}} (\sigma_{11} - \sigma_{33})\, dx_2 \tag{71}$$

If $\beta = 45°$, we have

$$\delta' \cos 2\varphi' = \tfrac{1}{2}n_0^3 \int_{x'_{20}}^{x'_{2*}} [(\pi_{11} - \pi_{12})(\tfrac{1}{2}\sigma_{11} + \tfrac{1}{2}\sigma_{22} - \sigma_{33}) + \pi_{44}\sigma_{12}]\, dx'_2 \tag{72}$$

We have assumed that the stress distribution is axisymmetric. Therefore, the integrals in eqns (71) and (72) depend only on the distance ξ of the ray from the cylinder axis and they do not depend on the direction of the ray. Subtracting eqn (72) from eqn (71) and expressing stresses in cylindrical coordinates yields

$$\frac{2}{n_0^3 R}[\delta(\xi)\cos 2\varphi(\xi) - \delta'(\xi)\cos 2\varphi'(\xi)]$$

$$= \int_0^{\sqrt{1-\xi^2}} (\sigma_r - \sigma_\theta)[(\pi_{11} - \pi_{12})\cos 2\theta - \pi_{44}\sin 2\theta]\, d\eta \tag{73}$$

Let us express $\sigma_r - \sigma_\theta$ as

$$\sigma_r - \sigma_\theta = \sum_{k=1}^{l} c_{2k}\rho^{2k} \tag{74}$$

Introducing the expansion of eqn (74) into eqn (73) and integrating gives

$$\frac{2}{n_0^3 R}[\delta(\xi)\cos 2\varphi(\xi) - \delta'(\xi)\cos 2\varphi'(\xi)] = \sum_{k=1}^{l} c_{2k} J_k(\xi) \tag{75}$$

where

$$J_k(\xi) = (\pi_{11} - \pi_{12})\left(2\xi I'_{k-1} - \frac{1}{\xi} I'_k\right) - \pi_{44}\frac{\xi}{k}(1 - \xi^{2k}) \tag{76}$$

and I'_k are determined by eqns (68) and (69).

Measuring $\delta(\xi)$, $\varphi(\xi)$, $\delta'(\xi)$ and $\varphi'(\xi)$ at m $(m \geqslant l)$ values of ξ we obtain a system of equations

$$\frac{2}{n_0^3 R}[\delta(\xi_i)\cos 2\varphi(\xi_i) - \delta'(\xi_i)\cos 2\varphi'(\xi_i)]$$

$$= \sum_{k=1}^{l} c_{2k} J_k(\xi_i) \qquad i = 1, 2, \ldots, m \tag{77}$$

from which the coefficients c_{2k} can be calculated by the method of least squares.

From the equilibrium equation

$$\frac{\partial \sigma_r}{\partial r} + \frac{\sigma_r - \sigma_\theta}{r} + \frac{\partial \sigma_{rz}}{\partial z} = 0 \tag{78}$$

we have, taking into account that at the boundary $\sigma_r = 0$,

$$\sigma_r(\rho) = \int_\rho^1 \frac{\sigma_r - \sigma_\theta}{\rho} \, d\rho + \int_\rho^1 \frac{\partial \sigma_{rz}}{\partial \zeta} \, d\rho \tag{79}$$

where

$$\zeta = \frac{z}{R} \tag{80}$$

Inserting eqns (63) and (74) into eqn (79) we obtain

$$\sigma_r(\rho) = \sum_{k=1}^{l} \left\{ \frac{c_{2k}}{2k}(1 - \rho^{2k}) + d'_{2k+1}\left[\tfrac{1}{2}(1 - \rho^2) - \frac{1}{2k+2}(1 - \rho^{2k+2})\right] \right\} \tag{81}$$

where

$$d'_{2k+1} = \frac{\partial d_{2k+1}}{\partial \zeta} \tag{82}$$

From eqns (74) and (81) we now have

$$\sigma_\theta(\rho) = \sum_{k=1}^{l} \left\{ c_{2k}\left[\frac{1}{2k} - \frac{2k+1}{2k}\rho^{2k}\right] + d'_{2k+1}\left[\tfrac{1}{2}(1-\rho^2)\right.\right.$$

$$\left.\left. - \frac{1}{2k+2}(1-\rho^{2k+2})\right]\right\} \tag{83}$$

Let us express σ_z again according to eqn (37). If we insert the expansion of eqn (37) into eqn (71) then, taking into account eqns (32) we obtain, after integration

$$\sum_{k=0}^{l} b_{2k} \frac{1}{\xi} I'_k(\xi_i) = \sum_{k=1}^{l} \left\{ c_{2k}\left[\frac{\xi_i}{2k}(\text{arc cos } \xi_i - I'_{k-1}(\xi_i))\right.\right.$$

$$+ \frac{1}{2k}\sqrt{(1-\xi_i^2)} - \frac{2k+1}{2k\xi_i} I'_k(\xi_i)$$

$$\left. + d'_{2k+1}\xi_i\left[\frac{2k+1}{2k+2}\text{arc cos } \xi_i - I'_0(\xi_i)\right]\right\}$$

$$- \frac{\delta(\xi_i)\cos 2\varphi(\xi_i)}{n_0^3(\pi_{11} - \pi_{12})R} \qquad (i = 1, 2, \ldots, m) \tag{84}$$

From the latter system of equations coefficients b_{2k} can be calculated. Thus, all the stress components have been determined.

5. EXPERIMENTS

5.1. Experimental Technique

For optical measurements the crystals were put into an immersion bath equipped with a device which permitted the crystal to be turned about its axis. For crystals of NaCl ($n_0 = 1.54$) and KCl ($n_0 = 1.49$) the immersion fluid was a mixture of medical vaseline oil ($n = 1.48$) and α-monobromonaphthalene ($n = 1.66$).

Characteristic angles, φ, φ', and integral optical path differences, δ, δ', were measured point-by-point in a polariscope KSP-5 which permits rotation of polariser and analyser synchronously. The polariscope is equipped with a coordinate device with a precision of 0.1 mm. The optical path differences were measured with the aid of a Krasnov compensator SKK-2. Figure 7 shows part of the polariscope with a crystal in the immersion bath.

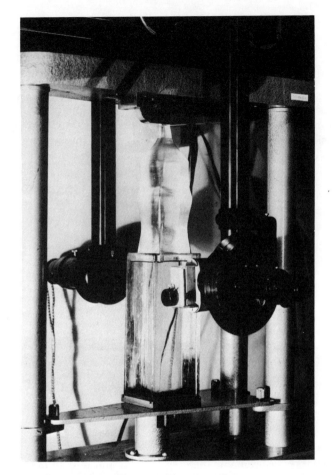

Fig. 7. The polariscope SKK-5 with a crystal in the immersion bath.

Measurements were carried out for all the cross-section and for further calculations the average for both halves of the cross-section was used. Practical calculations have shown that the number of measuring points should be about three times greater than the number of coefficients in the power series which determine the stress.

Since the problem of determining the stresses on the basis of experimental data is an ill-determined one, the number of terms in the power series should be kept to a reasonable minimum.

5.2. Stresses in the Middle Part of a NaCl Cylinder

The cylinder was 110 mm long and had a radius of 26 mm. In the middle part of the cylinder the parameter of the isoclinic relative to the cylinder axis was 0 over practically all of the cross-section. Therefore, the axial stress gradient could be ignored.

Integral optical path differences were measured scanning the cross-section once with the wave normal parallel to the [010] axis ($\beta = 0$), and for the second time with the wave normal parallel to the [$\bar{1}$10] axis ($\beta = 45°$, Fig. 3). The results are shown in Fig. 8. The symmetry of the curves δ and δ' is not ideal, but may be considered satisfactory.

Piezo-optic coefficients of NaCl are:[52]

$$\pi_{11} - \pi_{12} = -6 \cdot 7 \times 10^{-7} \, \text{MPa}^{-1}, \qquad \pi_{44} = -4 \cdot 5 \times 10^{-7} \, \text{MPa}^{-1}.$$

The experimental data were elaborated according to the algorithm described in Section 3. Stress distributions calculated from eqns (57)–(59), with $l = 4$, are shown in Fig. 9.

It can be seen that the distribution of the axial stress deviates considerably from a simple parabola. Such an anomalous stress distribution has been theoretically predicted by Reznikov.[30]

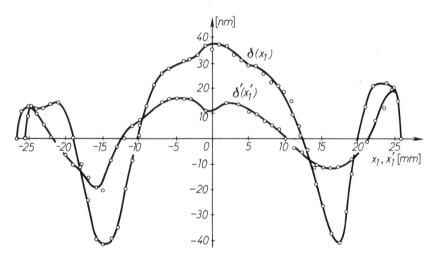

Fig. 8. Integral optical path differences in a cylindrical NaCl crystal: δ-wave normal parallel to the [010] direction, δ'-wave normal parallel to the [$\bar{1}$10] direction.

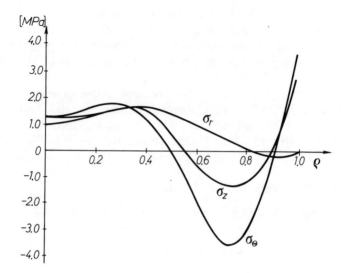

Fig. 9. Stress distribution in the cylindrical NaCl crystal.

The average axial stress over the cross-section is about 20% of its maximum value (at the boundary). Thus, the macrostatic axial equilibrium condition is not exactly fulfilled. The reason for this may either be experimental error or that not all the birefringence is due to elastic stresses.

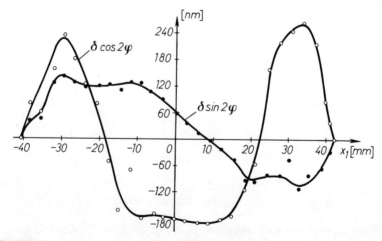

Fig. 10. Experimental data for a cylindrical KCl crystal, cross-section 1, wave normal in the [010] direction.

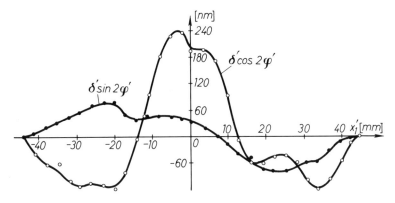

Fig. 11. As Fig. 10, with wave normal in $[\bar{1}10]$ direction.

5.3. Stresses in a KCl Cylinder with Axial Stress Gradient

The cylinder was 360 mm long and its radius was 45 mm. Measurements of the characteristic angles, φ and φ', as well as of the integral optical path differences, δ and δ', were carried out with the wave normal in the [010] and $[\bar{1}10]$ directions, in two cross-sections 1 and 2, 15 mm apart.

Experimental results are shown in Figs 10 to 13. Stresses were determined according to the algorithm described in Section 4.

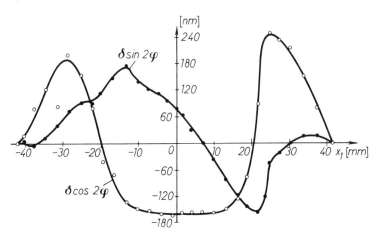

Fig. 12. As Fig. 10, with cross-section 2, wave normal in [010] direction.

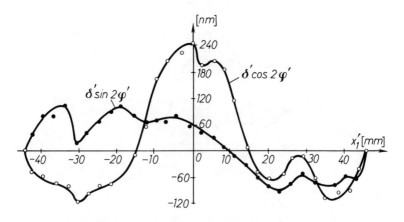

Fig. 13. As Fig. 11, with wave normal in [$\bar{1}$10] direction.

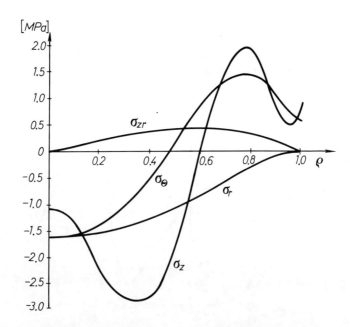

Fig. 14. Stress distribution in the cylindrical KCl crystal.

The piezo-optic coefficients of KCl are:[52]

$$\pi_{11} - \pi_{12} = 16 \cdot 4 \times 10^{-7} \, \text{MPa}^{-1}, \qquad \pi_{44} = -44 \cdot 2 \times 10^{-7} \, \text{MPa}^{-1}$$

The stress distribution is shown in Fig. 14. The average axial stress in this case is 11% of the maximum axial stress. Thus, the macrostatic equilibrium condition is fulfilled considerably better than in the former case.

6. CONCLUSIONS

In this chapter we have assumed that the stress distribution in the crystal is axisymmetric and that the birefringence is a homogeneous linear function of the stress components (eqn (6)). If, besides elastic stresses, the birefringence may also be due to plastic deformations, dislocations, etc., the optical properties of the crystal should be characterised by the distribution of the dielectric tensor. The possibilities of determining the latter will be considered in the following chapter.

The experimental technique considered above is very simple. The crystal is put in an immersion bath and viewed in a common polariscope. If no axial stress gradient is present, integral optical path differences are measured by scanning the cross-section. The wave normal should once be parallel and once form an angle of 45° to the crystallographic axis. The coefficients which determine the stress distribution are calculated from systems of linear equations. Tables of the functions G, F and I permit easy formation of the corresponding systems of equations if the measurements are carried out at values of ξ as given in the tables.

If the axial stress gradient is present, measurements are carried out for two adjacent cross-sections, the parameter of the isoclinic also being recorded.

Since optical path differences are usually small, a compensator should be used in the measurements.

It is important to point out that the method described above avoids use of the sum rule (eqn (45)). This is due to the fact that, in the case of a crystal, passing of light through the cross-section in different directions gives new information about the stress distribution. For the case of an isotropic axisymmetric body all directions in a cross-section are equivalent and application of the sum rule is inevitable.

The method described in this chapter has proved useful for checking the quality of single crystals as well as for elaborating a suitable growing technology for them.

REFERENCES

1. Pockels, F., *Lehrbuch der Kristalloptik*, Teubner-Verlag, Leipzig, 1906.
2. Mueller, H., Theory of the photoelastic effect of cubic crystals, *Phys. Rev.*, **47** (1935), 947–57.
3. Coker, E. G. and Filon, L. N. G., *A Treatise on Photoelasticity*, Cambridge Univ. Press, New York, 1957.
4. Ramachandran, G. N. and Ramaseshan, S., Crystal optics, In: *Encyclopedia of Physics*, Vol. 25/1, ed S. Flügge, Springer-Verlag, Berlin, 1961, pp. 1–217.
5. Krasnov, V. M., Determination of stress in cubic crystals by optical methods, *Uch. Zap. Len. Gos. Univ.*, **13** (87) (1944), 97–114. (In Russian.)
6. Krasnov, V. M. and Stepanov, A. V., Investigation of the beginning of fracture by optical methods, I, *Zh. Eksp. Teor. Fiz.*, **23** (2) (1952), 199–210. (In Russian.)
7. Krasnov, V. M. and Stepanov, A. V., Photoelastic investigation of stresses in an anisotropic plate under concentrated load, *Zh. Eksp. Teor. Fiz.*, **25** (1) (1953), 98–106. (In Russian.)
8. Goodman, L. E. and Sutherland, J. G., Elasto-plastic stress optical effect in silver chloride single crystals, *J. Appl. Phys.*, **24** (5) (1953), 577–84.
9. Krasnov, V. M., Stepanov, A. V. and Shvedko, E. F., Photoelastic investigation of stress in an anisotropic plate under concentrated load, *Zh. Eksp. Teor. Fiz.*, **34** (4) (1958), 894–8. (In Russian.)
10. Bugakov, I. I. and Grakh, I. I., Investigation of photoelasticity of anisotropic bodies, *Vest. Len. Gos. Univ.*, **19** (1968), 102–8. (In Russian.)
11. Zapletal, A., Fotoelasticita obecne anisotropni hmoty, *Stavebn. Casop.*, **21** (11) (1973), 817–29.
12. Bugakov, I. I., Grakh, I. I. and Konakova, N. S., Application of photoelasticity by determining mechanical stress in crystals, *Proc. Seventh All-Union Conf. on Photoelasticity*, Vol. 4, Tallinn, 1971, pp. 124–7. (In Russian.)
13. Afanas'ev, I. I. and Grakh, I. I., On photoelasticity of cubic crystals, *Proc. Seventh All-Union Conf. on Photoelasticity*, Vol. 4, Tallinn, 1971, pp. 134–40. (In Russian.)
14. Afanas'ev, I. I., An anisotropic analogue of the Wertheim law for stress determination in cubic crystals, *Proc. Eighth All-Union Conf. on Photoelasticity*, Vol. 4, Tallinn, 1979, pp. 17–22. (In Russian.)
15. Bond, W. L. and Andrus, J., Photographs of the stress field around edge dislocations, *Phys. Rev.*, **101** (3) (1956), 1211.
16. Indenbom, V. L. and Tomilovskii, G. E., Internal stresses around single dislocations, *Dokl. Akad. Nauk SSSR*, **115** (4) (1957), 723–6. (In Russian.)
17. Bullough, R., Birefringence caused by edge dislocation in silicon, *Phys. Rev.*, **115** (4) (1957), 723–6.

18. Indenbom, V. L. and Tomilovskii, T. E., Macroscopic edge dislocations in corund crystals, *Kristallografiya*, **2** (1) (1957), 190–2. (In Russian.)
19. Penning, P., Generation of imperfections in germanium crystals by thermal strain, *Philips Res. Repts*, **13** (1) (1958), 79–97.
20. Sakamoto, M., Quenching effect in KCl crystals, *J. Phys. Soc. Japan*, **14** (11) (1959), 1506–12.
21. Indenbom, V. L., Stresses and dislocations by crystal growth, *Izv. Akad. Nauk SSSR, Fiz.*, **37** (11) (1973), 2258–67. (In Russian.)
22. Indenbom, V. L. and Nikitenko, V. I., New applications and generalizations of the photoelasticity of crystals, *Proc. Eighth All-Union Conf. on Photoelasticity*, Vol. 4, Tallinn, 1979, pp. 46–57. (In Russian.)
23. Grechushnikov, B. N. and Brodovskii, D., Thermal stresses in cubic crystals, *Kristallografiya*, **1** (5) (1956), 597–9.
24. Sirotin, E. I., Thermal stresses which arise by heating and cooling of single crystals, *Kristallografiya*, **1** (6) (1956), 708–17.
25. Chebanova, T. S., On calculation of stress during crystal growth, *Mekh. Tverd. Tela*, (3) (1968), 63–72. (In Russian.)
26. Indenbom, V. L., Zhitomirskii, I. S. and Chebanova, T. S., Theoretical investigation of stresses which arise during the growing process, In *Crystal Growth*, Vol. 8, Akad. Nauk SSSR, Nauka, Moscow, 1968, pp. 303–9. (In Russian.)
27. Indenbom, V. L., Zhitomirskii, I. S., Morozovskaya, N. N. and Chebanova, T. S., Numerical determination of residual stresses which arise during growth of a semi-infinite cylinder, In *Thermal Stresses in Construction Elements*, Vol. 9, Naukova Dumka, Kiev, 1970, pp. 136–48. (In Russian.)
28. Vakhrameev, S. S., Calculation of thermal stresses in crystals grown from the melt, *Uch. Zap. Latv. Gos. Univ.*, **237** (1975), 101–22. (In Russian.)
29. Chernysheva, M. A., Measuring internal stresses in corund rods, In *Methods and Apparatus for Checking the Corund Crystals*, Nauka, Moscow, 1968, pp. 78–84. (In Russian.)
30. Reznikov, B. A., On anomalous residual stress distribution by thermal treatment of single crystals, *Fiz. Tverd. Tela*, **5** (9) (1963), 2526–9. (In Russian.)
31. Hornstra, J. and Penning, P., Birefringence due to residual stress in silicon, *Philips Res. Repts*, **14** (3) (1959), 237–49.
32. Nikitenko, V. I. and Indenbom, V. L., A comparison of stresses and dislocations in germanium crystals, *Kristallografiya*, **6** (6) (1961), 432–8. (In Russian.)
33. Josepson, J., Investigation of the stress distribution in a KCl single crystal, *Proc. Eighth All-Union Conf. on Photoelasticity*, Vol. 4, Tallinn, 1979, pp. 58–62. (In Russian.)
34. Konakova, N. S. and Lebedev, A. I., Application of the scattered light method for solving three dimensional problems of mechanics of anisotropic solids, *Proc. Eighth All-Union Conf. on Photoelasticity*, Vol. 4, Tallinn, 1979, pp. 69–72. (In Russian.)
35. Josepson, J., Stress distribution in cubic single crystals of cylindrical form by scattered-light photoelasticity, *ENSV Tead. Akad. Toim., Füüs., Mat.*, **28** (2) (1979), 146–50.

132   H. K. Aben

36. Aben, H., *Integrated Photoelasticity*, McGraw-Hill, New York, 1979.
37. Brosman, E. J., Aben, H. K. and Kaplan, M. S., A nondestructive method for checking residual stresses in cubic single crystals, In: *Physics and Chemistry of Crystals*, Kharkov, 1977, pp. 93–8. (In Russian.)
38. Brosman, E., On the application of integrated photoelasticity of cubic single crystals, *ENSV Tead. Akad. Toim.*, *Füüs.*, *Mat.*, **26** (4) (1977), 457–61. (In Russian.)
39. Brosman, E. J., Aben, H. K. and Kaplan, M. S., A nondestructive method for stress analysis in transparent cubic crystals, In: *Production and Investigation of Single Crystals*, (1) Kharkov, 1978, pp. 75–9.
40. Aben, H. and Brosman, E., Integrated photoelasticity of cubic single crystals, *VDI-Berichte*, **313** (1978), 45–50.
41. Aben, H. K. and Brosman, E. J., Determination of stress in cubic single crystals by integrated photoelasticity, *Theor. and Appl. Mech.* (*Bulg. Acad. Sci.*), **9** (1) (1978), 99–100. (In Russian.)
42. Aben, H. and Brosman, E., Integrated photoelasticity of cubic single crystals, *Lectures, Seventh Congress on Material Testing*, Vol. 2, Budapest, 1978, pp. 109–12.
43. Aben, H. K., Brosman, E. J. and Kaplan, M. S., Integrated photoelasticity of cubic single crystals, *Proc. Eighth All-Union Conf. on Photoelasticity*, Vol. 4, Tallinn, 1979, pp. 14–16. (In Russian.)
44. Brosman, E. J., Goriletskii, V. I., Kaplan, M. S., Lyakhov, V. V., Sumin, V. I. and Eidelman, L. G., Distribution of residual stresses in a KCl crystal grown from the melt, *Proc. Eighth All-Union Conf. on Photoelasticity*, Vol. 4, Tallinn, 1979, pp. 30–3. (In Russian.)
45. Tronko, V. D. and Golovach, G. P., Mueller and Jones matrices for a retardation plate with rotating fast axis, *Kristallografiya*, **18** (3) (1973), 459–64.
46. Doyle, J. F. and Danyluk, H. T., Integrated photoelasticity for axisymmetric problems, *Exp. Mech.*, **18** (6) (1978), 215–110.
47. Spencer, E. G., Denton, R. T., Bateman, T. B., Snow, W. and Van Uitert, L. G., Microwave elastic properties in nonmagnetic garnets, *J. Appl. Phys.*, **34** (10) (1963), 3059–60.
48. O'Rourke, R. C. and Saenz, A. W., Quenching stresses in transparent isotropic media and the photoelastic method, *Quart. Appl. Math.*, **8** (3) (1950), 303–11.
49. Poritsky, H., Analysis of thermal stresses in sealed cylinders and the effect of viscous flow during anneal, *Phys.*, **5** (12) (1934), 406–11.
50. Vest, C. M., Interferometry of strongly refracting axisymmetric phase objects, *Appl. Optics*, **14** (7) (1975), 1601–6.
51. Timoshenko, S. P. and Goodier, J. N., *Theory of Elasticity*, McGraw-Hill, New York, 1970.
52. Gorbach, S. S., Demtsova, L. A., Dobrzhanskii, G. F., Markovskii, B. J. and Shaskol'skaya, M. P., Piezo-optic coefficients of the crystals LiF, NaF, NaCl, NaI, KCl, KBr, KI and RbI, *Kristallografiya*, **14** (4) (1969) 729–32. (In Russian.)

4

ON THE INVESTIGATION OF AXISYMMETRIC DIELECTRIC TENSOR FIELDS BY INTEGRATED PHOTOELASTICITY

H. K. Aben and J. I. Josepson

Institute of Cybernetics, Academy of Sciences of the Estonian SSR, Tallinn, USSR.

ABSTRACT

While a great number of methods have been proposed and applied for the determination of axisymmetric dielectric constant scalar fields, little attention has been paid to the investigation of dielectric tensor fields, which is much more complicated. The latter problem becomes important in checking the quality of single crystals, fibres, plastic specimens etc., where birefringence cannot be related directly to elastic stresses and may have a more complicated origin.

In this chapter, integrated photoelastic techniques are applied for investigating several particular axisymmetric dielectric tensor fields. The theory of optical phenomena in an inhomogeneous dielectric medium is described briefly. Investigation of the dielectric tensor distribution in a plane of symmetry of an axisymmetric body is considered. The problem is simplified in the case of spherical symmetry. The case when the characteristic angle equals ±45° through the cross-section of a cylindrical specimen is also considered. Finally, determination of the birefringence and the twist angle in twisted fibres is described.

1. INTRODUCTION

In photoelasticity, the stress distribution in models is determined on the basis of measurements of the change in the polarisation of light when it passes through the model. By interpretation of the experimental results a linear relationship between the dielectric and stress tensors is postulated. Thus, by elaborating the experimental data the powerful apparatus of the mathematical theory of elasticity can be used (equations of equilibrium and of compatibility, macrostatic equilibrium conditions, etc.).

However, birefringence in transparent objects often has a more complicated origin. This may be due to inherent birefringence of polymer molecules, dislocations, plastic deformations, etc. In this case an attempt to interpret the experimental results as elastic stresses may lead to considerable errors.

The distribution of optical anisotropy may sometimes be of major importance and can determine the quality of the object, e.g. in the case of plastic scintillators, inhomogeneity of the distribution of the dielectric tensor worsens its optical properties while the internal stresses are, practically, of no importance. Distribution of the dielectric tensor also characterises the quality of single crystals, determines the type of spherulite, etc.

We may conclude that determination of dielectric tensor fields in transparent objects is a problem in its own right. We should mention that from the methodological point of view it is more correct in photoelasticity to first determine the dielectric tensor field in the model and only after this to introduce postulates about the relationship between dielectric and stress tensors and to determine the stress distribution.

There is a lot of literature devoted to the determination of the refractive index scalar field, i.e. to the case when the dielectric tensor is spherical. A review of the methods developed for that case is given in E. W. Marchand's book.[1]

Determination of the dielectric tensor field is much more complicated. Very simple particular cases have been considered in refs 2 and 3. The general theory of the optical phenomena for the case when polarised light passes through an inhomogeneous dielectric tensor field is given in ref. 4. Various possibilities of interpreting the experimental results while determining a three-dimensional dielectric tensor field in the general case are described,[5,6] with no practical applications.

Since determination of three-dimensional dielectric tensor fields is very complicated, it seems reasonable to start with a consideration of simple, particular cases. In this paper we investigate the possibilities of applying the usual integrated photoelasticity technique to obtain information about axisymmetric dielectric tensor fields. The general theory of the method is described and several particular cases with experimental examples are considered.

2. THEORY

Propagation of light in an inhomogeneous optically anisotropic medium is governed by the equations[4]

$$\frac{dB_1}{dz} = -\tfrac{1}{2}iC(\varepsilon_{11} - \varepsilon_{22})B_1 - iC\varepsilon_{12}B_2$$

$$\frac{dB_2}{dz} = -iC\varepsilon_{21}B_1 + \tfrac{1}{2}iC(\varepsilon_{11} - \varepsilon_{22})B_2 \tag{1}$$

where B_j denote transformed components of the electric vector which describe the polarisation of light in arbitrary coordinate axes, ε_{ij} is the dielectric tensor, z is the coordinate in the direction of light propagation, and

$$C = \frac{\omega}{2c\sqrt{\varepsilon}} \tag{2}$$

In the latter formula ε is a dielectric constant of the medium in the initial state.

In the case when the secondary principal directions of birefringence rotate along the wave normal it may be useful to apply secondary principal directions as coordinate axes. In this case eqns (1) take the form

$$\frac{dB_1}{dz} = -\tfrac{1}{2}iC(\varepsilon_1 - \varepsilon_2)B_1 + \frac{d\varphi}{dz}B_2$$

$$\frac{dB_2}{dz} = -\frac{d\varphi}{dz}B_1 + \tfrac{1}{2}iC(\varepsilon_1 - \varepsilon_2)B_2 \tag{3}$$

where $d\varphi/dz$ is the rotation of the secondary principal directions and ε_j are secondary principal values of the dielectric tensor.

It has been shown[4] that in the case of a linear optical medium the solution of eqns (1) and (3) can be written as

$$\begin{pmatrix} B_{1*} \\ B_{2*} \end{pmatrix} = U \begin{pmatrix} B_{10} \\ B_{20} \end{pmatrix} \tag{4}$$

where B_{j0} are components of the incident light vector, B_{j*} are components of the emergent light vector, and U is a two-by-two unitary matrix.

The most general expression for a two-by-two unitary unimodular matrix is[7]

$$U = \begin{pmatrix} e^{i\xi} \cos \theta & e^{i\zeta} \sin \theta \\ -e^{-i\zeta} \sin \theta & e^{-i\xi} \cos \theta \end{pmatrix} \tag{5}$$

where, in our case, ξ, ζ and θ are functions of the dielectric tensor distribution between the points of entrance and the emergence of light.

On the basis of eqns (4) and (5) it can be shown that for an inhomogeneous birefringent medium there always exist two perpendicular directions of the polariser by which the light emerging from the medium is also linearly polarised. Those directions of vibration of the incident light are named primary, and those of the emergent light, the secondary characteristic directions. They are determined by the angles α_0 and α_*, respectively

$$\tan 2\alpha_0 = \frac{\sin (\zeta + \xi) \sin 2\theta}{\sin 2\xi \cos^2 \theta - \sin 2\zeta \sin^2 \theta} \tag{6}$$

$$\tan 2\alpha_* = \frac{\sin (\zeta - \xi) \sin 2\theta}{\sin 2\xi \cos^2 \theta - \sin 2\zeta \sin^2 \theta} \tag{7}$$

Primary and secondary characteristic directions, as a rule, make an angle α, which is named the characteristic angle

$$\tan 2\alpha = \tan 2(\alpha_* - \alpha_0) = \frac{2 \sin 2\theta \cos \xi \cos \zeta}{\sin^2 \xi - \sin^2 \zeta - \cos 2\theta(\cos^2 \xi + \cos^2 \zeta)} \tag{8}$$

Phase retardation between light vibrations along characteristic directions is named the characteristic phase retardation Δ_*

$$\cos \Delta_* = \cos 2\xi \cos^2 \theta + \cos 2\zeta \sin^2 \theta \tag{9}$$

The primary and secondary characteristic directions and the characteristic phase retardation determine completely the transformation of the polarisation of light in an inhomogeneous medium. They are also the parameters which can be measured experimentally to determine the distribution of the dielectric tensor in the medium.

If the medium has a plane of symmetry perpendicular to the direction of light propagation, the characteristic angle vanishes ($\alpha_0 = \alpha_*$, $\alpha = 0$).

If there is no rotation of the secondary principal directions in the wave normal, i.e. if in eqns (3) $d\varphi/dz = 0$, the phase retardation Δ can be expressed as[4]

$$\Delta = C \int (\varepsilon_1 - \varepsilon_2)\, dz \qquad (10)$$

The latter equation is named the integral Wertheim law.

Often it is more comfortable to deal with principal refractive indices n_j, instead of the dielectric tensor; n_j can be expressed as

$$n_j = \sqrt{\varepsilon_j} \qquad (11)$$

For an optical path difference δ we get the following expression

$$\delta = \frac{\lambda}{2\pi}\Delta = \int (n_1 - n_2)\, dz \qquad (12)$$

By derivation of the latter formula we have assumed that the optical anisotropy of the medium is weak, i.e.

$$n_1^2 - n_2^2 \cong 2n_0(n_1 - n_2) \qquad (13)$$

where n_0 is the refractive index of the medium.

We should mention that the refractive index is not a tensor, although its dependence on direction is determined by the dielectric tensor.[8]

3. PLANE OF SYMMETRY

Let us pass a beam of polarised light through the plane of symmetry x–y of an axisymmetric body which is put into an immersion bath (Fig. 1). In this case $\varepsilon_{rz} = 0$ and there is no rotation of the secondary principal axes on a light ray. Therefore, the integral Wertheim law (eqns (10) and (12)) holds.

The optical properties of the cross-section are determined through the radial distribution of the three principal refractive indices n_r, n_θ and n_z. Equation (12) may be written in the form

$$\delta(x) = 2 \int_0^{y_*} (n_z - n_r \cos^2 \theta - n_\theta \sin^2 \theta)\, dy \qquad (14)$$

or after transformations, in dimensionless coordinates

$$\delta(\xi) = 2R \int_0^{\eta_*} [(n_z - n_\theta)\sin^2 \theta + (n_z - n_r)\cos^2 \theta]\, d\eta \qquad (15)$$

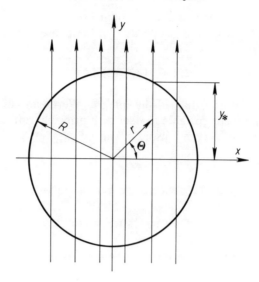

Fig. 1. Passing of light through the plane of symmetry of an axisymmetric body.

where

$$\xi = \frac{x}{R}, \qquad \eta = \frac{y}{R} \tag{16}$$

Equation (15) shows that the integral optical path difference is determined through the differences in the refractive indices, as expected. Let us express the latter as

$$n_z - n_\theta = \sum_{k=0}^{m} a_{2k}\rho^{2k} \tag{17}$$

$$n_z - n_r = \sum_{k=0}^{m} b_{2k}\rho^{2k} \tag{18}$$

where a_{2k} and b_{2k} are coefficients to be determined, and

$$\rho = \frac{r}{R} \tag{19}$$

Since at $\rho = 0$ we have $n_r = n_\theta$, it follows that $b_0 = a_0$.
Inserting eqns (17) and (18) into eqn (15), and integrating, yields

$$\frac{\delta(\xi)}{2R} = \sum_{k=0}^{m} [a_{2k}L_{2k}(\xi) + b_{2k}H_{2k}(\xi)] \tag{20}$$

where

$$L_0 = \sqrt{1 - \xi^2} - |\xi| \arccos |\xi|$$

$$L_2 = \tfrac{1}{3}\sqrt{1 - \xi^2}(1 - \xi^2)$$

$$L_4 = \tfrac{1}{5}\sqrt{1 - \xi^2}(1 - \tfrac{1}{3}\xi^2 - \tfrac{2}{3}\xi^4)$$

$$L_6 = \tfrac{1}{7}\sqrt{1 - \xi^2}(1 - \tfrac{1}{5}\xi^2 - \tfrac{4}{15}\xi^4 - \tfrac{8}{15}\xi^6)$$

(21)

and, in general

$$L_{2k} = \xi^{2k+1} \int_0^{\arccos \xi} \frac{\sin^2 \theta}{\cos^{2k+2} \theta} \, d\theta \tag{22}$$

The functions H_{2k} can be calculated recursively by the formula

$$H_{2k} = \frac{1}{2k - 1}[\xi^2\sqrt{1 - \xi^2} + 2(k - 1)\xi^2 H_{2k-2}] \tag{23}$$

with

$$H_0 = |\xi| \arccos |\xi|$$

$$H_2 = \xi^2\sqrt{1 - \xi^2}$$

$$H_4 = \tfrac{1}{3}\xi^2\sqrt{1 - \xi^2}(1 + 2\xi^2)$$

$$H_6 = \tfrac{1}{5}\xi^2\sqrt{1 - \xi^2}(1 + \tfrac{4}{3}\xi^2 + \tfrac{8}{3}\xi^4)$$

(24)

The first four functions, L_{2k} and H_{2k}, are shown in Figs 2 and 3, and are given numerically in Tables 1 and 2.

If the optical path difference $\delta(\xi)$ is measured at a number of values of ξ, eqn (20) formally gives a system of equations for coefficients a_{2k} and b_{2k}. However, the system in eqn (20) is ill-determined since two series of unknown coefficients a_{2k} and b_{2k} are related to only one series of experimental data $\delta(\xi)$. Therefore we lack experimental data to solve the problem.

Additional experimental data can be gained by the scattered light method. If we pass the light ray along the y-axis, the scattered light method permits the determination of the distribution of $n_z - n_\theta$, i.e. the coefficients a_{2k} in eqn (17). Now, in eqn (20) only the coefficients b_{2k} are unknown and can be calculated. As a matter of fact, the scattered light method permits the full determination of the distributions of

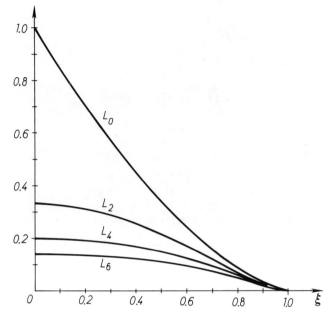

Fig. 2. Functions L_0, L_2, L_4 and L_6.

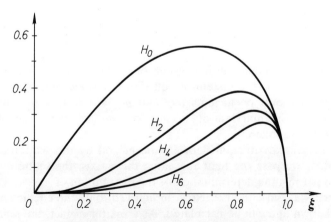

Fig. 3. Functions H_0, H_2, H_4 and H_6.

TABLE 1
Functions $L_{2k}(\xi)$

ξ	L_0	L_2	L_4	L_6
0·00	1·000	0·3333	0·2000	0·1429
0·05	0·9227	0·3321	0·1996	0·1426
0·10	0·8479	0·3283	0·1983	0·1419
0·15	0·7757	0·3221	0·1962	0·1406
0·20	0·7059	0·3135	0·1931	0·1388
0·25	0·6387	0·3026	0·1891	0·1364
0·30	0·5741	0·2894	0·1840	0·1335
0·35	0·5121	0·2740	0·1778	0·1299
0·40	0·4528	0·2566	0·1704	0·1256
0·45	0·3962	0·2374	0·1617	0·1204
0·50	0·3424	0·2165	0·1516	0·1144
0·55	0·2915	0·1942	0·1400	0·1047
0·60	0·2436	0·1707	0·1270	0·0993
0·65	0·1988	0·1463	0·1125	0·0899
0·70	0·1574	0·1214	0·0966	0·0791
0·75	0·1194	0·0965	0·0796	0·0669
0·80	0·0852	0·0720	0·0616	0·0534
0·85	0·0552	0·0487	0·0433	0·0388
0·90	0·0300	0·0276	0·0255	0·0236
0·95	0·0106	0·0101	0·0098	0·0094
1·00	0·0000	0·0000	0·0000	0·0000

$n_z - n_\theta$ and $n_z - n_r$. However, the scattered light method is time consuming, and due to non-uniform light scattering in a real specimen its precision is not high. Therefore in the case under consideration it is more effective to use the integral path difference $\delta(\xi)$ measurements that permit us to limit measurements by the scattered light method to the y-axis.

As an example, the distribution of the optical anisotropy in the middle cross-section of a cylindrical scintillator made of polymethylmethacrylate was determined by the combined integrated and scattered light method. The length of the cylinder was 200 mm, radius $R = 25$ mm.

The cylinder was investigated in an immersion bath in a mixture of α-bromonaphthalene and vaseline oil ($n_0 = 1·51$). The specimen was covered with a thin transparent coating to avoid the chemical influence of the immersion fluid.

TABLE 2

Functions $H_{2k}(\xi)$

ξ	H_0	H_2	H_4	H_6
0·00	0·0000	0·0000	0·0000	0·0000
0·05	0·0760	0·0025	0·0008	0·0005
0·10	0·1470	0·0099	0·0034	0·0020
0·15	0·2130	0·0222	0·0074	0·0046
0·20	0·2738	0·0392	0·0141	0·0082
0·25	0·3295	0·0605	0·0227	0·0132
0·30	0·3798	0·0859	0·0338	0·0196
0·35	0·4246	0·1148	0·0476	0·0276
0·40	0·4637	0·1466	0·0645	0·0376
0·45	0·4968	0·1808	0·0847	0·0499
0·50	0·5235	0·2165	0·1083	0·0650
0·55	0·5436	0·2526	0·1352	0·0832
0·60	0·5563	0·2880	0·1651	0·1052
0·65	0·5610	0·3210	0·1975	0·1310
0·70	0·5567	0·3499	0·2310	0·1605
0·75	0·5420	0·3721	0·2635	0·1930
0·80	0·5148	0·3840	0·2918	0·2262
0·85	0·4715	0·3806	0·3102	0·2554
0·90	0·4059	0·3531	0·3083	0·2704
0·95	0·3016	0·2818	0·2635	0·2466
1·00	0·0000	0·0000	0·0000	0·0000

The integral path differences (Fig. 4) were measured in a polariscope KSP-5 with the Krasnov compensator SKK-2. The parameter of the isoclinic was practically 0° over all the cross-section.

The scattered light measurements were carried out on an original photoelectric scattered light polariscope constructed by J. Josepson. In Fig. 5 the cumulative phase retardation measurement results are given when light passes through the model along the y-axis (Fig. 1). These measurements permitted the distribution of $n_z - n_\theta$ (coefficients a_{2k}) to be determined, which is shown in Fig. 6. After that, coefficients b_{2k} were calculated from eqn (20). The distribution of $n_z - n_r$ is also shown in Fig. 6.

To check the results the distribution of $n_z - n_r$ was also determined by the method of scattered light by scanning the cross-section parallel to the y-axis and measuring the derivative of the optical path difference on the x-axis. The results are shown in Fig. 6.

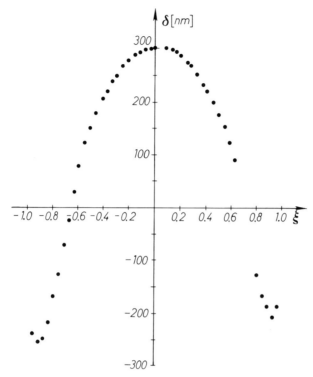

Fig. 4. The distribution of the integral optical path differences in the cylindrical plastic scintillator.

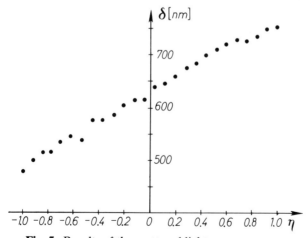

Fig. 5. Results of the scattered light measurements.

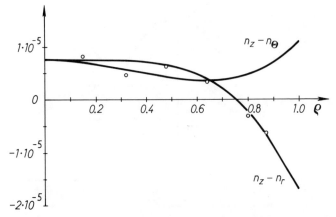

Fig. 6. Distribution of the optical anisotropy in the middle cross-section of the cylindrical plastic scintillator, determined by the combined method; small circles give the results of the scattered light method.

4. THE CASE $n_r = n_\theta$

It is not always possible to use the scattered light method in addition to integral optical measurements. The integral optical measurements are sufficient to determine the distribution of the optical anisotropy if we may assume that $n_r = n_\theta$. This assumption may be approximately valid if we investigate a comparatively long specimen in which the optical anisotropy is caused mostly by axial deformations.

If $n_r = n_\theta$, eqn (15) yields

$$\delta(\xi) = 2R \int_0^{\eta*} (n_z - n_r)\, d\eta \tag{25}$$

and, instead of eqn (20), we have

$$\frac{\delta(\xi)}{2R} = \sum_{k=0}^{m} a_{2k} G_{2k}(\xi) \tag{26}$$

where

$$G_0 = \sqrt{1-\xi^2}$$

$$G_2 = \tfrac{1}{3}\sqrt{1-\xi^2}(1+2\xi^2)$$

$$G_4 = \tfrac{1}{5}\sqrt{1-\xi^2}(1+\tfrac{4}{3}\xi^2+\tfrac{8}{3}\xi^4)$$

$$G_6 = \tfrac{1}{7}\sqrt{1-\xi^2}(1+\tfrac{6}{5}\xi^2+\tfrac{8}{5}\xi^4+\tfrac{16}{5}\xi^6)$$

$$\tag{27}$$

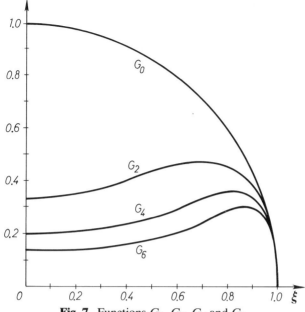

Fig. 7. Functions G_0, G_2, G_4 and G_6.

TABLE 3

Functions G_{2k} (ξ)

ξ	G_0	G_2	G_4	G_6
0·00	1·0000	0·3333	0·2000	0·1429
0·05	0·9987	0·3346	0·2004	0·1431
0·10	0·9950	0·3383	0·2017	0·1439
0·15	0·9887	0·3444	0·2039	0·1452
0·20	0·9798	0·3527	0·2072	0·1471
0·25	0·9682	0·3631	0·2118	0·1497
0·30	0·9539	0·3752	0·2178	0·1531
0·35	0·9367	0·3888	0·2254	0·1575
0·40	0·9165	0·4033	0·2349	0·1631
0·45	0·8930	0·4182	0·2464	0·1703
0·50	0·8660	0·4330	0·2598	0·1794
0·55	0·8352	0·4468	0·2752	0·1907
0·60	0·8000	0·4587	0·2921	0·2044
0·65	0·7599	0·4674	0·3100	0·2208
0·70	0·7141	0·4713	0·3276	0·2396
0·75	0·6614	0·4685	0·3431	0·2599
0·80	0·6000	0·4560	0·3535	0·2796
0·85	0·5268	0·4293	0·3535	0·2942
0·90	0·4359	0·3807	0·3339	0·2941
0·95	0·3122	0·2920	0·2732	0·2560
1·00	0·0000	0·0000	0·0000	0·0000

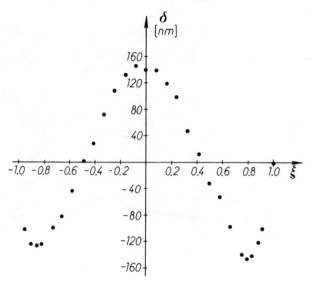

Fig. 8. Distribution of integral optical path difference in a long glass cylinder.

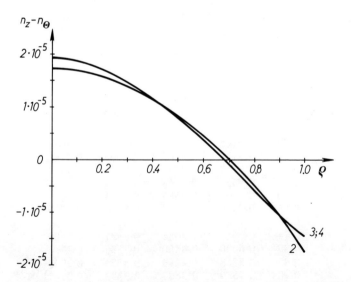

Fig. 9. Distribution of $n_z - n_r$ in a long glass cylinder; numbers denote the number of terms in eqn (26).

and the following functions $G_{2k}(\xi)$ can be calculated recursively by the aid of the formula

$$G_{2k} = \frac{1}{2k+1}[\sqrt{1-\xi^2}+2k\xi^2 G_{2k-2}] \tag{28}$$

Functions G_0, G_2, G_4 and G_6 are shown graphically in Fig. 7 and are given numerically in Table 3. To determine the coefficients a_{2k} in eqn (17) integral optical path differences should be measured for l ($l \geqslant m$) values of ξ. Equation (26) now yields l equations from which the coefficients a_{2k} can be determined by the method of least squares.

In Fig. 8 the distribution of integral optical path differences in the cross-section of a long glass cylinder with $R = 12$ mm is shown.[9] The experimental data were elaborated according to the algorithm described above. In Fig. 9 the distribution of $n_z - n_r$ is shown for various numbers of terms in eqn (17). Interpretation of the same experimental data in stresses is given in ref. 9.

5. SPHERICAL SYMMETRY

If polarised light is passed through a diametral section of a sphere, we have $n_z = n_\theta$. In this case eqn (15) yields

$$\delta(\xi) = 2R \int_0^{\eta_*} (n_\theta - n_r)\cos^2\theta \, d\eta \tag{29}$$

Putting expansion

$$n_\theta - n_r = \sum_{k=1}^{m} b_{2k}\rho^{2k} \tag{30}$$

into eqn (29) and integrating gives

$$\frac{\delta(\xi)}{2R} = \sum_{k=1}^{m} b_{2k}H_{2k}(\xi) \tag{31}$$

Measuring the integral optical path difference $\delta(\xi)$ for l values of ξ ($l \geqslant m$) permits the coefficients b_{2k} to be calculated from eqn (31) by the method of least squares. This method is more general than that described in ref. 3, and permits birefringence in spherulites to be determined without additional assumptions.

6. THE CASE $\alpha_0 = 45°$

By investigating the optical properties of plastic scintillators as well as of YAG crystals, on several occasions, the authors have observed the following phenomenon. Through almost the whole cross-section of the cylindrical specimen the characteristic angle (parameter of the isoclinic) α_0 equals $\pm 45°$, while the integral phase retardation $\delta(0)$ on the axis is practically zero. A possible explanation of this phenomenon is as follows.

Assume that the axes of the indicatrix in the plane $z - r$ form with the r-axis an angle of $\pm 45°$. Since $\alpha_0 = \pm 45°$, practically throughout the cross-section, our assumption is valid at the boundary of the cross-section and should approximately hold elsewhere also.

Since, by $\xi = 0$, we have $\delta = 0$, we may conclude that the cross-section of the indicatrix, perpendicular to the r-axis, is a circle.

According to these assumptions the dielectric tensor can be expressed as (Fig. 10)

$$(\varepsilon_{ij}) = \begin{pmatrix} \dfrac{\varepsilon_1 + \varepsilon_3}{2} & 0 & \dfrac{\varepsilon_1 - \varepsilon_3}{2} \\ 0 & \varepsilon_2 & 0 \\ \dfrac{\varepsilon_1 - \varepsilon_3}{2} & 0 & \dfrac{\varepsilon_1 + \varepsilon_3}{2} \end{pmatrix} \tag{32}$$

Let us rotate this tensor about the z-axis for an angle θ

$$(\varepsilon'_{ij}) = (\theta)^{-1}(\varepsilon_{ij})(\theta) \tag{33}$$

where

$$(\theta) = \begin{pmatrix} \cos \theta & -\sin \theta & 0 \\ \sin \theta & \cos \theta & 0 \\ 0 & 0 & 1 \end{pmatrix} \tag{34}$$

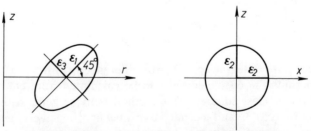

Fig. 10. Dielectric ellipsoid for the case $\alpha_0 = 45°$.

Here we have used a formalism analogous to that used in refs 2 and 3.

Taking into account eqns (32) and (34), eqn (33) yields

$$\varepsilon'_{xx} = \frac{\varepsilon_1 + \varepsilon_3}{2} \cos^2 \theta + \varepsilon_2 \sin^2 \theta$$

$$\varepsilon'_{xy} = \varepsilon'_{yx} = \left(-\frac{\varepsilon_1 + \varepsilon_3}{2} + \varepsilon_2 \right) \sin \theta \cos \theta$$

$$\varepsilon'_{xz} = \varepsilon'_{zx} = \frac{\varepsilon_1 - \varepsilon_3}{2} \cos \theta \tag{35}$$

$$\varepsilon'_{yy} = \frac{\varepsilon_1 + \varepsilon_3}{2} \sin^2 \theta + \varepsilon_2 \cos^2 \theta$$

$$\varepsilon'_{yz} = \varepsilon'_{zy} = -\frac{\varepsilon_1 - \varepsilon_3}{2} \sin \theta$$

$$\varepsilon'_{zz} = \frac{\varepsilon_1 + \varepsilon_3}{2}$$

The polarisation of light is determined through the components of the tensor ε_{ij} which are in the plane $x-z$. In this plane the principal directions of birefringence are determined by the relationship

$$\tan 2\varphi = \frac{2\varepsilon'_{zx}}{\varepsilon'_{xx} - \varepsilon'_{zz}} = \frac{2 \cos \theta}{(2k + 1) \sin^2 \theta} \tag{36}$$

and birefringence Δn by

$$\Delta n = (n_3 - n_1) \sqrt{\tfrac{1}{4} \sin^4 \theta (2k + 1)^2 + \cos^2 \theta} \tag{37}$$

where

$$k = \frac{n_1 - n_2}{n_3 - n_1} \tag{38}$$

From eqn (36) it follows that if $k = -\frac{1}{2}$, then $\varphi \equiv 45°$. Since this holds true for every layer in the model, there is no rotation of the principal axes of birefringence on a light ray and we have $\alpha_0 = \varphi = 45°$. This is exactly what we have observed experimentally. Therefore, we may assume that $k = -\frac{1}{2}$.

Following eqns (12) and (37), the integral optical path difference can be expressed as

$$\delta(\xi) = 2R \int_0^{\arccos \xi} (n_3 - n_1) \cos \theta \, d\eta \tag{39}$$

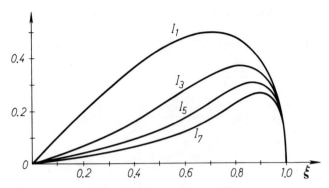

Fig. 11. Functions I_1, I_3, I_5 and I_7.

TABLE 4
Functions $I_{2k+1}(\xi)$

ξ	I_1	I_3	I_5	I_7
0·00	0·0000	0·0000	0·0000	0·0000
0·05	0·0499	0·0167	0·0100	0·0072
0·10	0·0994	0·0338	0·0202	0·0144
0·15	0·1483	0·0517	0·0306	0·0218
0·20	0·1960	0·0705	0·0414	0·0294
0·25	0·2421	0·0908	0·0530	0·0374
0·30	0·2862	0·1126	0·0653	0·0459
0·35	0·3279	0·1361	0·0789	0·0551
0·40	0·3666	0·1613	0·0940	0·0653
0·45	0·4019	0·1882	0·1109	0·0767
0·50	0·4330	0·2165	0·1299	0·0897
0·55	0·4593	0·2457	0·1513	0·1049
0·60	0·4800	0·2752	0·1753	0·1227
0·65	0·4940	0·3038	0·2015	0·1435
0·70	0·4999	0·3299	0·2293	0·1677
0·75	0·4961	0·3514	0·2573	0·1949
0·80	0·4800	0·3648	0·2828	0·2237
0·85	0·4478	0·3649	0·3005	0·2501
0·90	0·3923	0·3426	0·3005	0·2647
0·95	0·2966	0·2774	0·2596	0·2432
1·00	0·0000	0·0000	0·0000	0·0000
0·96	0·2688	0·2548	0·2416	0·2292
0·97	0·2358	0·2265	0·2174	0·2092
0·98	0·1950	0·1899	0·1849	0·1801
0·99	0·1397	0·1378	0·1360	0·1342

Let us express $n_3 - n_1$ in the form

$$n_3 - n_1 = \sum_{k=0}^{m} d_{2k+1} \rho^{2k+1} \tag{40}$$

Inserting eqn (40) into eqn (39) and integrating yields

$$\frac{\delta(\xi)}{2R} = \sum_{k=0}^{m} d_{2k+1} I_{2k+1}(\xi) \tag{41}$$

where

$$I_1 = \xi\sqrt{1-\xi^2}$$

$$I_3 = \tfrac{1}{3}\xi\sqrt{1-\xi^2}(1+2\xi^2)$$

$$I_5 = \tfrac{1}{5}\xi\sqrt{1-\xi^2}(1+\tfrac{4}{3}\xi^2+\tfrac{8}{3}\xi^4)$$

$$I_7 = \tfrac{1}{7}\xi\sqrt{1-\xi^2}(1+\tfrac{6}{5}\xi^2+\tfrac{24}{15}\xi^4+\tfrac{48}{15}\xi^6)$$

$$\tag{42}$$

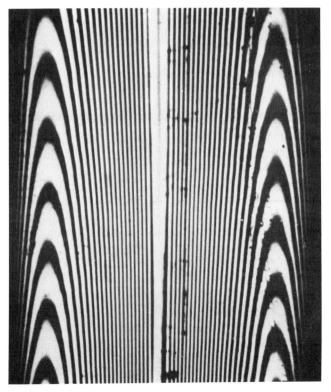

Fig. 12. Integral fringe pattern of a cylindrical plastic scintillator.

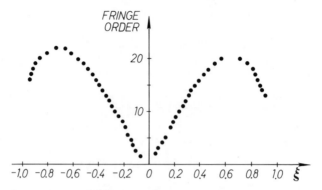

Fig. 13. Distribution of integral fringe order in the plastic cylindrical scintillator.

and the following functions, I_{2k+1}, may be calculated recursively from

$$I_{2k+1} = \frac{1}{2k+1} \xi\sqrt{1-\xi^2} + \frac{2k}{2k+1} \xi^2 I_{2k-1} \qquad (43)$$

The first four functions I_{2k+1} are shown in Fig. 11 and are given numerically in Table 4.

As an example, the integral fringe pattern of a cylindrical plastic scintillator, where the phenomenon $\alpha_0 = \pm 45°$ was observed, is shown in Fig. 12. In Fig. 13 the distribution of the integral fringe order in one

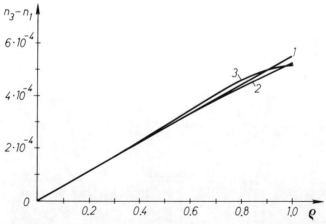

Fig. 14. Radial distribution of $n_3 - n_1$; numbers denote the number of terms in eqn (40).

cross-section is shown. The experimental data were elaborated according to the algorithm described above. In Fig. 14 the radial distribution of $n_3 - n_1$ is shown.

7. TWISTED FIBRES

In fibre optics the following problem often arises. In a drawn monofilament, the long-chain molecules tend to be oriented along the length of the filament. When the monofilament is twisted, the molecules will spiral around the filament axis. On a diameter of the cross-section, the orientation of the molecules rotates continuously, being in the centre of the cross-section parallel to the fibre axis (Fig. 15). The properties of the monofilament are determined[10,11] by the total angle of twist, φ, as well as by the birefringence Δn.

We have the case when the principal directions of birefringence rotate uniformly through the specimen while the birefringence is constant. This case has been considered in detail with the aid of the theory of characteristic directions.[4,12]

Let us denote the total phase retardation when the light passes the fibre along its diameter, ignoring rotation of the principal axes, by Δ

$$\Delta = \frac{4\pi}{\lambda} R \, \Delta n \tag{44}$$

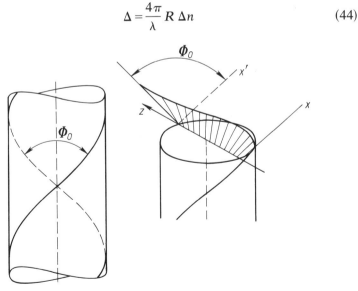

Fig. 15. A twisted fibre.

We also use the following notations

$$K = \frac{2\varphi}{\Delta} \qquad (45)$$

$$S = \sqrt{1 + K^2} \qquad (46)$$

$$\eta = \tfrac{1}{2} S \, \Delta \qquad (47)$$

In this case eqns (8) and (9) give the following expressions for the characteristic angle α and for the characteristic phase retardation Δ_*

$$\tan \alpha = \frac{\tan \varphi - \dfrac{K}{S} \tan \eta}{1 + \dfrac{K}{S} \tan \varphi \tan \eta} \qquad (48)$$

$$\cos \Delta_* = 1 - \frac{2}{S^2} \sin^2 \eta \qquad (49)$$

If, by passing light through the cross-section of the fibre along its diameter, the characteristic angle α and the characteristic phase retardation Δ_* are measured, the twist angle φ and parameter η, which determine the birefringence, can be calculated from the non-linear system of eqns (48) and (49). The values of φ and Δ can easily be determined on the basis of experimentally measured α and Δ_* with the aid of a nomogram.[4,12]

8. CONCLUSIONS

It has been shown that the theory of characteristic directions, successfully applied in three-dimensional photoelasticity, can also be used for investigating dielectric tensor fields. However, in comparison with photoelasticity the latter problem is more complicated since by interpreting the experimental results we cannot use the relationships which hold for the stress field. Therefore, only some simple, particular cases have been considered in this chapter.

It is opportune to notice that image reconstruction from projections, tomography,[13] has up to now dealt only with scalar fields. Therefore, this chapter can be considered as one of the first steps towards tensor field tomography.

It is obvious that integrated photoelasticity is only one possibility to investigate dielectric tensor fields. Interferometry, holography, shadow methods, etc., can also be used. Additional information can also be gained by measuring the bending of light rays in the specimen.[14,15] The methods described above have been used for checking the optical quality of plastic scintillators and single crystals of cylindrical form. A more general approach to the problem is under study.

REFERENCES

1. Marchand, E. W., *Gradient Index Optics*, Academic Press, New York, 1978.
2. Keith, H. D. and Padden, F. J., The optical behaviour of spherulites in crystalline polymers. Part I. Calculation of theoretical extinction patterns in spherulites with twisting crystalline orientation. *J. Polymer Sci.*, **39** (1959), 101–22.
3. Ong, C., Yoon, D. Y. and Stein, S., Correct measurement of birefringence in a nonuniform medium. *J. Polymer Sci., Polymer Phys. Ed.*, **12** (1974), 1319–25.
4. Aben, H., *Integrated Photoelasticity*, McGraw-Hill, New York, 1979.
5. Kubo, H. and Nagata, R., Determination of dielectric tensor fields in weakly inhomogeneous anisotropic media. *J. Opt. Soc. Amer.*, **69** (4) (1979), 604–10.
6. Kubo, H. and Nagata, R., Determination of dielectric tensor fields in weakly inhomogeneous anisotropic media. II. *J. Opt. Soc. Amer.*, **71** (3) (1981) 327–33.
7. Bhagavantam, S. and Venkatarayudu, T., *Theory of Groups and Its Application to Physical Problems*, Andhra University, Waltair, 1951.
8. Nye, J. F., *Physical Properties of Crystals*, Clarendon Press, Oxford, 1957.
9. Brosman, E., Determination of thermal stresses in cylinders by integrated photoelasticity. *Eesti NSV Tead. Akad. Toim., Füüs., Mat.*, **25** (4) (1976), 418–21. (In Russian.)
10. Astle-Fletcher, M. W., A method for evaluating the twist in nylon monofilaments. *J. Textile Inst.*, **48** (1957), T128–T132.
11. Desai, J. N. and Patel, R. M., Optics of a twisted birefringent monofilament. *Indian J. Pure Appl. Phys.*, **4** (11) (1966), 436–8.
12. Aben, H. K., The optical investigation of twisted fibres. *J. Textile Inst.*, **59** (11) (1968), 523–7.
13. Herman, G. T., *Image Reconstruction from Projections. The Fundamentals of Computerized Tomography*, Academic Press, New York, 1980.
14. Aben, H., Krasnowski, B., and Pindera, J., On nonrectilinear light propagation in integrated photoelasticity of axisymmetric bodies, *Eesti NSV Tead. Akad. Toim., Füüs., Mat.*, **31** (1), (1982), 65–73. (In Russian.)
15. Aben, H. K., Integrated photoelasticity of axisymmetric bodies. *Optical Engineering*, **21** (4) (1982), 689–95.

5

PHOTOELASTIC ANALYSIS OF COMPOSITE MATERIALS WITH STRESS CONCENTRATORS

E. E. Gdoutos

School of Engineering, Democritus University of Thrace, Xanthi, Greece

ABSTRACT

The optical method of photoelasticity is used for the determination of the singular stress field around stress concentrators in composite materials. The cases of a bimaterial plate with a crack along or perpendicular to the interface and a general shaped inclusion with cuspidal points along its boundary embedded in an elastic matrix are considered. In all these cases, the stress field distribution in the local neighbourhood of the point of stress singularity is used for the determination of the isochromatic fringe patterns. Generally, these patterns are governed by two parameters which for the case of cracks are known as the opening-mode and sliding-mode stress intensity factors. The general characteristic properties of the isochromatic patterns are analysed and discussed. For each case particular methods are developed for the determination of the stress field around the singular points by taking information from the isochromatic patterns.

1. INTRODUCTION

In recent years composite materials have gained popularity in engineering applications due to their flexibility in obtaining the desired mechanical and physical properties in combination with light-weight components. Two types of composites may generally be distinguished:

multi-layered composites made by bonding different materials and composites made up of a soft matrix and stiff filler particles.

For multi-layered composites, the strength of the bonds between the various separate layers is of particular importance for their structural integrity. The influence of defects such as manufacturing cavities developed during bonding or the casting process, or the flaws resulting from residual stresses in welded materials must be taken into consideration when estimating the strength of the composite. The determination of the stress field developed around these defects idealised as cracks is of special interest and constitutes the first step for calculation of the strength of bonds. On the other hand, for composites made up of a soft matrix and filler particles these particles are usually assumed to have a simple geometrical shape, such as spheroidal, ellipsoidal or cylindrical, which facilitates the analysis. However, in many composites, the reinforcing constituents are of irregular shape with sharp angles, like the various inorganic fillers, the metal or boron filaments, or the aggregate or sand particles in concrete. In such cases, in the sharp angle corners, high stress concentrations develop and act as nuclei for the generation of cracks and slip bands, leading to failure.

A great deal of effort has been spent in recent years on the determination of the stress field around defects in composite materials. Most of the works published are theoretical and are therefore limited by the necessary idealisations. It appears that experimental stress analysis methods such as photoelasticity, moiré, holography and interferometry have not made much progress in the solution of such problems. Recently some progress has been made at the Democritus University of Thrace in Greece for the application of the photoelastic method to the investigation of problems of composite materials having stress concentrators.[1-6] It is the purpose of the present work to report some of the preliminary results. In the following the special problems arising in the photoelastic determination of the stress field around cracks in bimaterial plates and inclusions of general shape with cuspidal points are analysed and discussed. The cases of a bimaterial plate with a crack along or perpendicular to the interface and rigid inclusions with cuspidal points embedded in a matrix are considered.

2. THE BIMATERIAL PLATE WITH A CRACK ALONG THE INTERFACE

In this section the isochromatic fringe patterns developed in the vicinity of the tip of a crack lying along the interface of two bonded dissimilar materials are studied. The case when the plate is subjected to a uniform uniaxial stress perpendicular to the crack plane is first considered to establish the characteristic properties of these patterns and their dependence on the elastic properties of both materials of the plate. Finally, a method for the determination of the stress intensity factors governing the stress field in the vicinity of the crack tip for any type of loading is developed.

2.1. The Stress Field

Consider two homogeneous, isotropic elastic materials which occupy the upper ($y > 0$) and lower ($y < 0$) half-planes with shear modulus of elasticity $G_{1,2}$ and Poisson's ratio $\nu_{1,2}$. The materials are bonded along the x-axis except for a finite segment of length $2a$ which forms an internal crack. The crack faces are free of loads and the composite plate is subjected to stresses at infinity.

This problem has been considered by Rice and Sih[7] and Erdogan[8] who gave the stress components σ_x, σ_y, τ_{xy} in the vicinity of the crack tip. Thus, the stress σ_{1x} for medium 1 ($0 < \theta < 180°$) is expressed by:

$$
\sigma_{1x} = \frac{K_I}{2\sqrt{2\pi r}} \left\{ \frac{1}{2} e^{-\beta(\pi-\theta)} \left[5 \cos\left(\frac{\theta}{2} + \beta \log\frac{r}{a}\right) + \cos\left(\frac{5\theta}{2} + \beta \log\frac{r}{a}\right) \right. \right.
$$
$$
\left. \left. + 4\beta \sin\theta \cos\left(\frac{3\theta}{2} + \beta \log\frac{r}{a}\right) \right] - e^{\beta(\pi-\theta)} \cos\left(\frac{\theta}{2} - \beta \log\frac{r}{a}\right) \right\}
$$
$$
- \frac{K_{II}}{2\sqrt{2\pi r}} \left\{ \frac{1}{2} e^{-\beta(\pi-\theta)} \left[5 \sin\left(\frac{\theta}{2} + \beta \log\frac{r}{a}\right) + \sin\left(\frac{5\theta}{2} + \beta \log\frac{r}{a}\right) \right. \right.
$$
$$
\left. \left. + 4\beta \sin\theta \sin\left(\frac{3\theta}{2} + \beta \log\frac{r}{a}\right) \right] + e^{\beta(\pi-\theta)} \sin\left(\frac{\theta}{2} - \beta \log\frac{r}{a}\right) \right\} \quad (1)
$$

and similar expressions valid for the other stresses.

In this relation (r, θ) are the polar coordinates of the point considered and β is a bimaterial constant given by

$$
\beta = \frac{1}{2\pi} \log \beta_0 \quad (2)
$$

with

$$
\beta_0 = \frac{G_1 + \kappa_1 G_2}{G_2 + \kappa_2 G_1} \quad (3)
$$

where $\kappa_i = 3 - 4\nu_i$ for plane strain or $\kappa_i = (3-\nu_i)/(1+\nu_i)$ for plane stress. The K_I and K_{II} stress intensity factors are independent of the coordinates (r, θ) but depend on the elastic constants of the two media and the loading conditions of the plate.

From the above definition of the bimaterial constant β_0 it is established that $\frac{1}{3} \leqslant \beta_0 \leqslant 3$. The values $\beta_0 = \frac{1}{3}$ and 3 correspond to $(G_2/G_1) = 0$ and ∞ respectively with $\kappa_{1,2} = 3$ (corresponding to the value of the Poisson's ratio $\nu = 0$ under plane strain conditions). As is shown from eqn (3), when the materials 1 and 2 are interchanged the constant β_0 takes the value $(1/\beta_0)$.

When both materials have the same elastic properties the bimaterial constant β is equal to zero and eqn (1) coincides with the value of σ_x for the case of a crack in an isotropic plate.

2.2. Isochromatic Patterns

The isochromatic pattern developed at the tip of the crack of the bimaterial composite loaded by a remote uniaxial stress σ normal to the crack has been studied. The stress intensity factors K_I and K_{II} for this case are given by:[7]

$$K_I = \frac{\cos (\beta \log 2a) + 2\beta \sin (\beta \log 2a)}{\cosh \pi \beta} \sigma \sqrt{\pi a} \qquad (4a)$$

$$K_{II} = -\frac{\sin (\beta \log 2a) - 2\beta \cos (\beta \log 2a)}{\cosh \pi \beta} \sigma \sqrt{\pi a} \qquad (4b)$$

The order N of the isochromatic fringes is given according to the photoelastic law by

$$\frac{Nf}{t} = \sigma_1 - \sigma_2 = [(\sigma_x - \sigma_y)^2 + 4\tau_{xy}^2]^{1/2} \qquad (5)$$

where σ_1, σ_2 are the principal stresses, f is the stress-optical constant of the material considered and t is the thickness of the plate.

Refer all lengths of the plate to the half-crack length, i.e. put $a = 1$. If the values of the stress components $\sigma_{1x}, \sigma_{1y}, \tau_{1xy}$ are introduced into eqn (5) for each value of the bimaterial constant β_0 at each point (r, θ) of the plate the corresponding value of the quantity $(Nf/\sigma t)$ is obtained. Thus, computer-generated isochromatic fringes around the crack tip in material 1 are obtained. Similarly, the isochromatic pattern in material 2 is derived. The study of isochromatic patterns can only be made for the values of β_0 lying in the interval $(1 \leqslant \beta_0 \leqslant 3)$, since the remaining values of β_0 are obtained by interchanging materials 1 and 2.

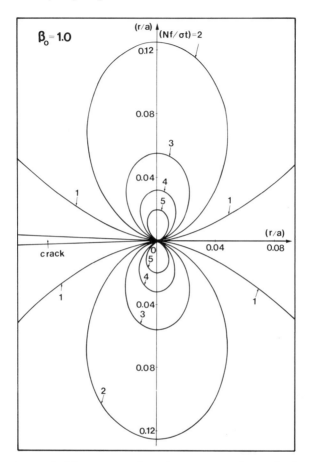

Fig. 1. Isochromatic fringe pattern for a homogeneous isotropic plate ($\beta_0 = 1$) subjected to a uniform stress normal to the crack axis.

Figures 1–3 present the isochromatic patterns for $\beta_0 = 1\cdot0$, $1\cdot2$ and $2\cdot0$. The photoelastic constant f and the material thickness t assumed the same value in both media 1 and 2. The value $\beta_0 = 1$ corresponds to the trivial case when both materials are the same and the isochromatic pattern has its well-known form.[9] It is observed from Fig. 1 that all fringes pass through the crack tip and they are symmetric with respect to both the crack axis and the axis passing from the crack tip and perpendicular to the crack axis. From Figs 2 and 3 with $\beta \neq 1$ we see

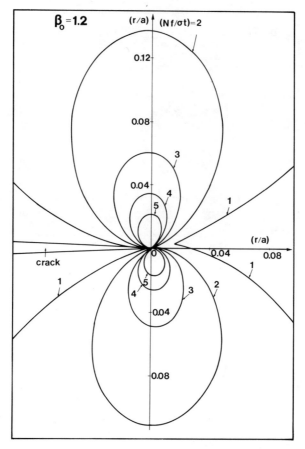

Fig. 2. Isochromatic fringe pattern for a bimaterial plate with $\beta_0 = 1\cdot2$ subjected to a uniform stress normal to the crack axis.

that the fringes no longer pass through the crack tip and are no longer symmetric with respect to the above two axes. Fringes of the same order, however, intersect the crack ligament at the same point. A counterclockwise rotation of the fringes close to the crack tip which increases with β_0 is observed. Also, there is a tendency for the fringes to pass through the crack tip and for fringes of the same order to present a corner point. Both these characteristics are predominant for small values of β_0 and almost disappear as β_0 increases. All Figs 1–3 correspond to the case when both materials of the composite have the

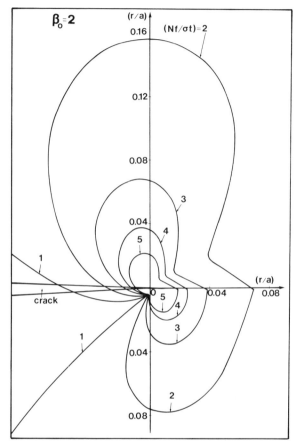

Fig. 3. Isochromatic fringe pattern for a bimaterial plate with $\beta_0 = 2$ subjected to a uniform stress normal to the crack axis.

same value of the photoelastic constant. For the general case with f_1 and f_2, the values of f for materials 1 and 2, the values of the quantity $(Nf/\sigma t)$ in the figures should be multiplied by (f_1/f) and (f_2/f) to obtain the actual isochromatic patterns. This results in a discontinuity of the same order fringes along the interface of the two media.

2.3. Determination of Stress Intensity Factors
The isochromatic fringe pattern around the crack tip depends on the values of stress intensity factors K_I and K_{II} and therefore it can be

used for their determination. At this point it should be emphasised that
the above analysis is valid only in the close vicinity of the crack tip and
therefore measurements on the isochromatic pattern should be taken
in this area. However, such measurements are influenced by the
following factors which alter the real meaning of the isochromatic
pattern: the radius of the slit which in the experiments simulates the
crack, the stress singularity resulting in an enhanced isochromatic-
fringe density so that the order of fringes may exceed the linear limit in
the stress-fringe curve of the material, and the triaxiality of the stress
field. While these characteristics are common to all cracks in the
problem under study further care should be taken to avoid measurements in an area surrounding the crack tip where the stress field
presents an oscillatory behaviour dictated by the elastic solution of the
problem (eqn (1)). However, such an area is very small and in most
cases it lies inside the region influenced by the aforementioned factors.
Thus, in general, the oscillatory behaviour of the stress field in the case
of the bimaterial composite does not introduce additional problems.

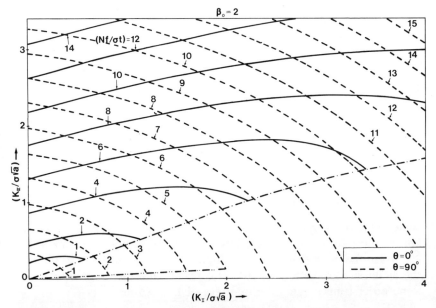

Fig. 4. Nomogram for the determination of K_I and K_{II} stress intensity factors.
The continuous curves correspond to the point ($\theta = 0°$, $r = 0.03$), the dotted
curves to the point ($\theta = 90°$, $r = 0.03$). The bimaterial constant β_0 takes the
value $\beta_0 = 2.0$.

On the basis of the above arguments and taking the radius r of the region in which measurements are made in the interval $(0 \cdot 01, 0 \cdot 1)$ nomograms were constructed for the immediate determination of K_I and K_{II}. The idea is to determine from the experiment the order N of the isochromatic at two points and calculate K_I and K_{II} from the solution of the corresponding two equations. For convenience the two points were obtained along the crack ligament $(\theta = 0)$ and the normal from the crack tip to the crack axis lying in material 1 $(\theta = 90°)$ and at a distance $(r/a) = 0 \cdot 03$ from the crack tip. Figure 4 presents such a nomogram for $\beta_0 = 2 \cdot 0$. In the horizontal and vertical axes of the figure the quantities $(K_I/\sigma\sqrt{a})$ and $(K_{II}/\sigma\sqrt{a})$ are drawn respectively, while the curves correspond to the normalised values of the fringe order $(Nf/\sigma t)$. The continuous lines correspond to $\theta = 0°$, the dotted lines to $\theta = 90°$. The use of this nomogram for the determination of K_I and K_{II} is simple. The order of the isochromatics N_1 and N_2 at the points $(r = 0 \cdot 03, \theta = 0°)$ and $(r = 0 \cdot 03, \theta = 90°)$ is measured. The thus determined values of $(Nf/\sigma t)$ define a point in the figure to which a definite pair of values of $(K_I/\sigma\sqrt{a})$ and $(K_{II}/\sigma\sqrt{a})$ correspond. It can be observed that the two groups of curves of the figure intersect each other and therefore the experimentally determined values of N_1 and N_2 define a point in the figure with high accuracy.

3. THE BIMATERIAL PLATE WITH A CRACK PERPENDICULAR TO THE INTERFACE

3.1. General Considerations

Bimaterial specimens of this type have recently been introduced into studies aiming to characterise the dynamic fracture behaviour of metals. Usually, these specimens consist of a starter section made of a brittle material with a crack and an arrest section made of a tougher material. The two sections are bonded along the common interface. An extensive study of crack arrest and reinitiation phenomena in such bimaterial specimens was made by Dally and Kobayashi.[10] They used dynamic photoelasticity to determine the position of the crack tip and the opening-mode stress intensity factor K during crack propagation. Theocaris and Milios[11] determined the crack velocity and the stress intensity factor during crack propagation using the method of caustics. In all these studies K was determined by the well-known singular solution of a crack in an isotropic medium. The main characteristic feature of this solution is that it presents an inverse square root singularity. This solution, however, cannot be used for the case when the crack meets the bimaterial interface. Indeed, in this case neither the stress singularity nor the angular stress distribution at the crack tip are the same as in the case of the isotropic material.[12] Thus, the value of the stress singularity is no longer equal to $0·5$ but depends on the modulus of elasticity and Poisson's ratio of the two materials of the composite. Analysis of the isochromatic patterns for the case when the crack meets the bimaterial interface was recently provided by the author.[1,2,6]

3.2. The Stress Distribution

Consider a bimaterial plate consisting of materials 1 and 2 and a crack in material 1 terminating perpendicularly at the interface. For symmetric applied loading the singular stress field in the vicinity of the crack tip can be expressed in the form[12]

$$\sigma_{ij} = \frac{K}{\sqrt{2}r^g} f_{ij}(\theta) \qquad (i, j = r, \theta) \tag{6}$$

where σ_{ij} are the polar components of stress, K is the stress intensity factor, g is the value of the stress singularity $(0 < g < 1)$ and $f_{ij}(\theta)$ are bounded functions.

The value of g is independent of the loading of the plate and depends only on the elastic constants of the two materials of the composite. The value of $\lambda = (1 - g)$ is the root with the smaller real part of the following equation

$$2\alpha \cos \pi\lambda - (\beta\lambda^2 + \gamma) = 0 \tag{7}$$

with

$$\alpha = (m + \kappa_2)(1 + m\kappa_1) \qquad \beta = -4(m + \kappa_2)(1 - m)$$

$$\gamma = (1 - m)(m + \kappa_2) + (1 + m\kappa_1)(m + \kappa_2) - m(1 + \kappa_1)(1 + m\kappa_1) \tag{8}$$

$$m = G_2/G_1$$

It can be shown from eqn (7) that λ is always real.

3.3. Isochromatic Patterns

Using the stress components from eqn (6) the isochromatic pattern for an aluminium-epoxy plate was obtained for the cases where the crack exists in the epoxy or the aluminium material; g takes the values

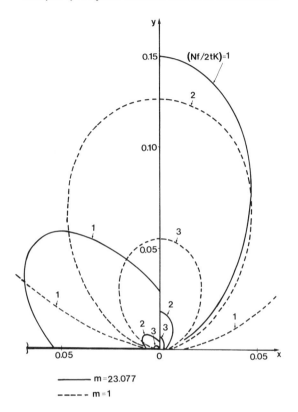

Fig. 5. Isochromatic fringe patterns for an epoxy-aluminium duplex specimen with the crack lying in the epoxy phase. The corresponding pattern for a plain specimen is indicated by dotted lines. (——) $m = 23.077$; (————) $m = 1$.

0.3381 and 0.8258 for these cases respectively. The patterns thus obtained are shown in Figs 5 and 6 for the cases when the crack exists in the epoxy and aluminium materials, respectively. In the same figures the corresponding patterns for a cracked isotropic plate are shown by dotted lines. A substantial difference between the two patterns is observed. The main characteristic of the figures in the bimaterial plate is that they are discontinuous along the interface. It has to be pointed out that in Figs 5 and 6 the stress-optical constant f and the thickness t for both materials assumed the same value.

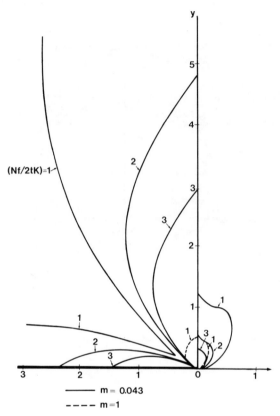

Fig. 6. As for Fig. 5 but with the crack lying in the aluminium phase. (——)
$m = 0.043$; (– – – –) $m = 1$.

3.4. Determination of Stress Intensity Factor

The stress intensity factor K, describing the singular behaviour of the
stress field in the vicinity of the crack tip can be determined from the
obtained isochromatic pattern in a similar manner as in the previous
case of the composite plate with a crack along the interface. In doing
so the singular stress components given from eqn (6) should be used.
In order to obtain an idea of the error introduced by calculating K
from the well-known formula of isotropic materials, Fig. 7 presents the
variation of $\Lambda = K/K_0$ versus the radial distance r for the polar angles
$\theta = 15°$, $30°$, $60° \sim 90°$, $120°$ and $150°$ for the case of Fig. 5. K_0 is the

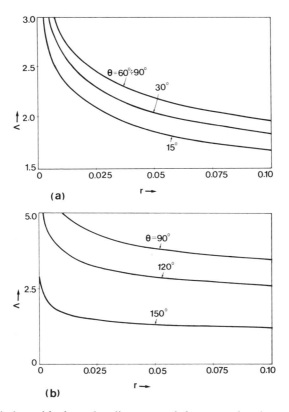

Fig. 7. Variation with the polar distance r of the correction factor Λ for the photoelastic determination of the stress intensity factor in an epoxy-aluminium duplex specimen with the crack lying in the epoxy phase.

stress intensity factor determined from the formula of isotropic materials. It is observed that Λ is always greater than unity and decreases with the radial distance r. Thus, the photoelastic formula for determining K in isotropic materials always underestimates K. We also see that for the higher toughness material $(0 < \theta < 90°)$, Λ increases with the polar angle θ (Fig. 7a), while the contrary happens for the brittle material $(90° < \theta < 180°)$ (Fig. 7b). This suggests that the error introduced in the determination of K by the photoelastic formula of isotropic materials is smaller for large distances r from the crack tip and small polar angles θ for the tougher material or greater polar angles for the brittle material. However, in the photoelastic determination of K data are usually obtained from the close vicinity of the crack tip, where the singular solution better describes the state of affairs. Otherwise, higher order terms are needed for the description of the isochromatic fringe pattern. For more information on these developments the interested reader is referred to refs 1, 2 and 6.

4. RIGID INCLUSIONS WITH CUSPIDAL POINTS

4.1. The Stress Field

Consider a rigid inclusion with cuspidal points embedded in an isotropic plate (Fig. 8). A reference frame of polar coordinates with origin at the cuspidal point 0 and x-axis coinciding with the tangent of the inclusion at 0 is attached to the plate. For this situation the polar components of stress σ_r, σ_θ, $\tau_{r\theta}$ in the vicinity of the point 0 are given by the following relations[13]

$$\sigma_r = \frac{1}{4\sqrt{2r}} \left[k_1 \left[5 \cos \frac{\theta}{2} + (2\kappa + 1) \cos \frac{3\theta}{2} \right] \right.$$
$$\left. - k_2 \left[5 \sin \frac{\theta}{2} + (2\kappa - 1) \sin \frac{3\theta}{2} \right] \right] \tag{9a}$$

$$\sigma_\theta = \frac{1}{4\sqrt{2r}} \left[k_1 \left[3 \cos \frac{\theta}{2} - (2\kappa + 1) \cos \frac{3\theta}{2} \right] \right.$$
$$\left. - k_2 \left[3 \sin \frac{\theta}{2} - (2\kappa - 1) \sin \frac{3\theta}{2} \right] \right] \tag{9b}$$

$$\tau_{r\theta} = \frac{1}{4\sqrt{2r}} \left[k_1 \left[\sin \frac{\theta}{2} - (2\kappa + 1) \sin \frac{3\theta}{2} \right] \right.$$
$$\left. + k_2 \left[\cos \frac{\theta}{2} - (2\kappa - 1) \cos \frac{3\theta}{2} \right] \right] \tag{9c}$$

It is observed from eqns (9a)–(9c) that the stress field in the vicinity of the cuspidal point presents an inverse square root singularity as in the case of cracks. The coefficients k_1 and k_2 are independent of the coordinates (r, θ) and depend only on the geometry of the inclusion, the applied stress and the Poisson's ratio of the plate. Due to the similarity of eqns (9) with the corresponding equations for a crack in a mixed mode stress field the coefficients k_1 and k_2 will be called the symmetric and skew-symmetric stress intensity factors respectively.

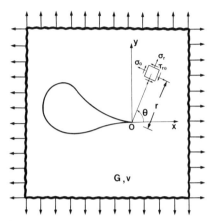

Fig. 8. Geometry of a general shaped inclusion with a cuspidal point embedded in an elastic plate.

If the above expressions for the stresses are introduced into the photoelastic law (eqn (5)), the equation of the isochromatic fringes in the local neighbourhood of the cuspidal point is obtained. Thus, after lengthy calculations, the following equation is obtained for the radius r of the isochromatic fringe of order N

$$r = \frac{1}{2}\left(\frac{t}{Nf}\right)^2 (a_{11}k_1^2 + 2a_{12}k_1k_2 + a_{22}k_2^2) \tag{10}$$

with

$$a_{11} = (1+\kappa)^2 - (1+2\kappa)\sin^2\theta$$

$$a_{12} = -\kappa\sin 2\theta \tag{11}$$

$$a_{22} = (1-\kappa)^2 + (2\kappa - 1)\sin^2\theta$$

In order to obtain an idea of the form of isochromatic patterns the cases of the astroidal, hypocycloidal and rectilinear inclusions are considered next.

4.2. The Astroidal, Hypocycloidal and Rectilinear Inclusions
For the astroidal inclusion the stress intensity factors k_1 and k_2 of eqn (9) are given by[13]

$$k_1^{(j)} = \frac{\sqrt{3a}}{4\kappa}\, p\left[\frac{\kappa - 1}{2} + \frac{3\kappa}{3\kappa + 1}\cos(\pi j - 2\beta)\right] \tag{12a}$$

$$k_2^{(j)} = \frac{\sqrt{3a}}{4}\, p\,\frac{3}{3\kappa - 1}\sin(\pi j - 2\beta) \tag{12b}$$

with $j = 0, 1, 2$ and 3 for the four cuspidal points of the inclusion. β defines the inclination angle of the inclusion relative to the loading direction.

Introducing these expressions of k_1 and k_2 into eqns (10) and (11) the values of the non-dimensional quantity (rN^2f^2/t^2p^2a) for each of the four cuspidal points of the inclusion are obtained. Figure 9 presents the isochromatic pattern thus obtained for the cuspidal points $j = 0(0_0)$ and $j = 3(0_3)$. The angle β and the material constant κ were given the values 30^0 and $1\cdot8$. It is observed that the isochromatics form closed loops which do not pass from the cuspidal points as occurs in the case of a crack in an isotropic plate.

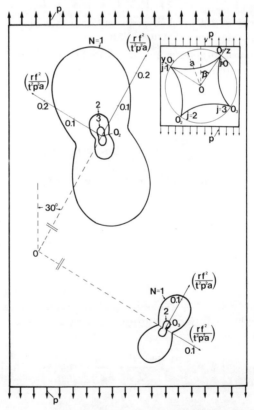

Fig. 9. Isochromatic patterns developed at the cuspidal points $j = 0(0_0)$ and $j = 3(0_3)$ of an astroidal inclusion for $\beta = 30°$ and $\kappa = 1\cdot8$.

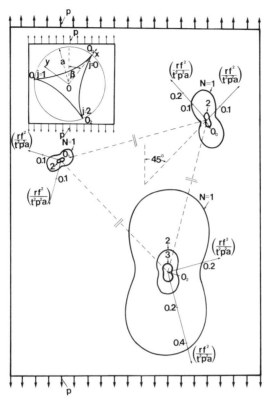

Fig. 10. Isochromatic patterns developed at the cuspidal points $j = 0(0_0)$, $j = 1(0_1)$ and $j = 2(0_2)$ of a hypocycloidal inclusion for $\beta = 45°$ and $\kappa = 1·8$.

For the hypocycloidal inclusion (Fig. 10) the stress intensity factors k_1 and k_2 are given by

$$k_1^{(j)} = \frac{\sqrt{2a}}{3\kappa} p \left[\frac{\kappa - 1}{2} + \cos \left(\frac{4\pi j}{3} - 2\beta \right) \right] \qquad (13a)$$

$$k_2^{(j)} = \frac{\sqrt{2a}}{3\kappa} p \sin \left(\frac{4\pi j}{3} - 2\beta \right) \qquad (13b)$$

Working as in the previous case the isochromatic patterns around all three cuspidal points of the inclusion were determined. Figure 10 presents these patterns for $\beta = 45°$ and $\kappa = 1·8$.

Finally, for the rigid rectilinear inclusion of length $2l$, inclined at an angle β to the direction of the applied stress σ, the values of k_1 and k_2 are given by

$$k_1 = -\frac{p\sqrt{l}}{2\kappa}\left(\frac{\kappa-1}{2}+\cos 2\beta\right) \tag{14a}$$

$$k_2 = -\frac{p\sqrt{l}}{2\kappa}\sin 2\beta \tag{14b}$$

The isochromatic pattern for $\beta = 60°$ and $\kappa = 1\cdot8$ is presented in Fig. 11.

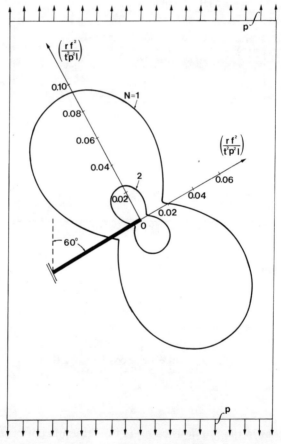

Fig. 11. Isochromatic fringe pattern around the end point of a fibre embedded in an infinite plate. The plate is subjected to a uniform uniaxial stress p at an angle $\beta = 60°$ with the fibre axis. The material constant κ takes the value $1\cdot8$.

4.3. Determination of Stress Intensity Factors k_1, k_2

In the above cases for the particular geometries of the inclusion and the type of loading considered the stress intensity factors are known and the photoelastic analysis was provided for the purpose of obtaining an idea of the form of isochromatics around cuspidal points of rigid inclusions. However, in general, in cases of complicated geometrical configurations of the inclusions and applied loads the analytical computation of stress intensity factors becomes difficult and therefore their experimental determination is of particular importance. Photoelasticity may provide a powerful tool for this purpose.

One easily measurable quantity in the isochromatic pattern is the angle θ_m that the tangent of the inclusion at the cuspidal point subtends with the line from the cuspidal point to the farthest point on a given loop. The condition for determining the position on an isochromatic loop, which has the largest distance from the cuspidal point, is expressed by the following relation:

$$\frac{\partial \tau_{max}}{\partial \theta} = 0 \qquad (15)$$

which, by taking into account eqns (10) and (11), leads to the following equation:

$$\frac{2\kappa - 1}{2\kappa + 1} \left(\frac{k_2}{k_1}\right)^2 - \frac{4\kappa}{2\kappa + 1} \left(\frac{k_2}{k_1}\right) \cot 2\theta_m - 1 = 0 \qquad (16)$$

Equation (16) for $\kappa = -1$ gives the well-known relation for the case of a crack in a mixed-mode stress field (eqn (15), ref. 14).

The variation of the ratio (k_2/k_1) versus the angle θ_m for various values of κ as expressed in eqn (16) was plotted in Fig. 12. This figure enables the direct determination of the ratio (k_2/k_1) by measuring the angle θ_m on the isochromatic loops. If the determined ratio (k_2/k_1) is now introduced into eqns (10) and (11) the value of k_1 can be determined from one measurement on the isochromatic pattern.

Determination of k_1 and k_2 can also be made by applying eqns (10) and (11) to two points of the isochromatic pattern with known values of the fringe order N. The values of k_1 and k_2 are calculated from the solution of the derived system of two equations with unknowns k_1 and k_2. It is evident that the measurements on the isochromatic pattern must be made in the close vicinity of the cuspidal point since, as we recede from it, eqn (9) and therefore eqns (10) and (11) no longer hold.

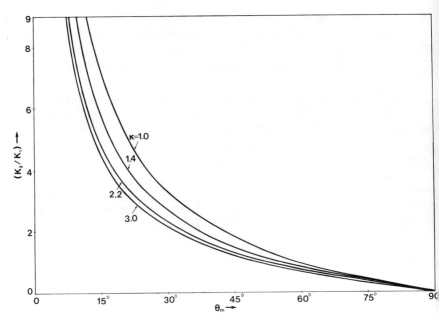

Fig. 12. Variation of the ratio (k_2/k_1) of the stress intensity factors k_1 and k_2 versus the angle θ_m that the tangent at the cuspidal point subtends with the line connecting the cuspidal point with the farthest point on a given loop. The diagram is valid for any form of geometry and loading of the plate. The material constant κ takes the values 1·0, 1·4, 2·2 and 3·0.

In order to facilitate the solution of the above system of equations, nomograms were constructed for the immediate determination of k_1 and k_2. The points at which the fringe order on the isochromatic loops is measured were taken along the x and y axes. Figure 13 presents such a nomogram for $\kappa = 2$. In the horizontal and vertical axes of this figure the stress intensity factors k_1 and k_2 are depicted, while the curves correspond to the values of the quantity (rN^2f^2/t^2). The continuous lines correspond to the x axis $(\theta = 0)$, while the dotted lines correspond to the y axis $(\theta = 90°)$. The use of this nomogram for the determination of k_1 and k_2 is simple. The order N of the isochromatic pattern at two points along the x and y axes is measured from experiment and then the corresponding values of the quantity (rN^2f^2/t^2) are calculated. The determined values of this quantity define a point in the figure to which a definite pair of values of k_1 and k_2 correspond. It can be seen that this point in the plane $k_1 - k_2$ is defined with adequate accuracy.

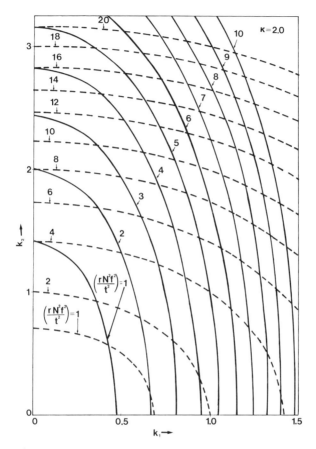

Fig. 13. Nomogram for the determination of k_1 and k_2 stress intensity factors. The continuous and dotted curves correspond to measurements on the isochromatic pattern along the x and y axes, respectively. The material constant κ takes the value 2·0.

5. CONCLUDING REMARKS

Although the method of photoelasticity has been used extensively for the determination of the singular stress field in the vicinity of crack tips in isotropic, homogeneous, elastic bodies not much progress has been made for the case of composites. In the present communication an attempt is made for the first time to use the isochromatic pattern of

E. E. Gdoutos

photoelasticity to analyse the stress field in the following cases usually encountered in engineering applications: (i) a bimaterial plate with a crack along or perpendicular to the interface, and (ii) a general shaped inclusion with cuspidal points along its boundary embedded in an elastic matrix. In all these problems attention is focused in the vicinity of the crack tip or the cuspidal point of the inclusion.

For the crack problems considered the singular behaviour of the stress field in the vicinity of the crack tip is not of the familiar form used in fracture mechanics. Thus, when the crack lies along a bimaterial interface the singular stress field is characterised by a logarithmic type of singularity, the stress field is of mixed-mode even for loads symmetric with respect to the crack plane and the angular dependence of the stress field is different than for a crack in an isotropic medium. When the crack meets perpendicularly a bimaterial interface with the crack tip at the interface the order of singularity is different from $-0\cdot5$ and the angular dependence of the stresses is not the same as in cracks in isotropic bodies. Finally, in the case of an inclusion with cuspidal points although the singularity is of the order of $-0\cdot5$, the stress field is dependent on the Poisson's ratio of the matrix material.

The form and the main characteristic properties of the isochromatic fringe patterns developed near the singular points in the above cases were studied and their peculiar character was analysed and discussed. It was found that these patterns are strongly dependent on the mechanical properties of the constituent materials of the composite and that they deviate more from the pattern of a crack in an isotropic plate the more the elastic constants of the two media differ. It was established that measurements on the isochromatic fringes enable the determination of the characteristic quantities of the singular stress field. Suitable nomograms facilitating this procedure were constructed. From the study it was concluded that the photoelastic method of stress analysis can successfully be used for analysing singular stress fields around stress concentrators in composites in an analogous manner to the case of cracks in isotropic, homogeneous bodies. For the latter case there is extensive evidence in the published literature.

ACKNOWLEDGEMENTS

The author is indebted to Professor G. Holister for his interest and his helpful discussions of the present communication during the author's visit to the Open University in September 1983, made possible through a grant from the British Council. The support of the Council and the interest in the author's visit of Mr J. Potts, Director of the British Council in Thessaloniki, is gratefully acknowledged.

REFERENCES

1. Gdoutos, E. E., On the photoelastic determination of K_I stress intensity factors in duplex specimens, *International J. Fracture*, **16** (1980), 776–7.
2. Gdoutos, E. E., Determination of stress intensity factors during crack arrest in duplex specimens, *International J. Solids and Structures*, **17** (1981), 683–5.
3. Gdoutos, E. E., Isochromatic patterns in a plate with a rigid fiber inclusion, *Fibre Science and Technology*, **15** (1981) 299–311.
4. Gdoutos, E. E., Photoelastic analysis of the stress field around singular points of rigid inclusions, *J. Applied Mechanics*, **49** (1982) 236–8.
5. Gdoutos, E. E. and Papakaliatakis, G., Photoelastic study of a bimaterial plate with a crack along the interface, *Engineering Fracture Mechanics*, **16** (1982) 77–87.
6. Gdoutos, E. E., Isochromatic fringe patterns in duplex specimens, *Fibre Science and Technology*, **16** (1982) 67–77.
7. Rice, J. R. and Sih, G. C., Plane problems of cracks in dissimilar media, *J. Applied Mechanics*, **32** (1965) 418–23.
8. Erdogan, F., Stress distribution in bonded dissimilar materials with cracks. *J. Applied Mechanics*, **32** (1965) 403–10.
9. Theocaris, P. S. and Gdoutos, E. E., A photoelastic determination of K_I stress intensity factors, *Engineering Fracture Mechanics*, **7** (1975), 331–9.
10. Dally, J. W. and Kobayashi, T., Crack arrest in duplex specimens, *International J. Solids and Structures*, **14** (1978), 121–9.
11. Theocaris, P. S. and Milios, J., Dynamic crack propagation in composites, *International J. Fracture*, **16** (1980) 31–41.
12. Cook, T. S. and Erdogan, F., Stresses in bonded materials with a crack perpendicular to the interface, *International J. Engineering Sciences*, **10** (1982) 677–97.
13. Panasyuk, V. V., Berezhnitskii, L. T. and Trush, I. I., Stress distribution about defects such as rigid sharp-angled inclusions, *Problemy Prochnosti*, **7** (1972) 3–9.
14. Gdoutos, E. E. and Theocaris, P. S., A photoelastic determination of mixed-mode stress-intensity factors, *Experimental Mechanics*, **18** (1978), 87–96.

6

PHOTOELASTIC STUDY OF CRACK PROBLEMS

E. E. Gdoutos

School of Engineering, Democritus University of Thrace, Xanthi, Greece

ABSTRACT

The problem of the photoelastic determination of the stress field around cracks under static and dynamic loading is considered. The direct application of photoelasticity to the solution of such problems necessitates particular attention to the evaluation of the experimental results. This mainly arises from the existing singularity at the tip of the crack resulting in a rapid variation of the stress components near to the crack tip. Thus particular methods and techniques for the determination of the crack tip stress intensity factors were developed. The use of a Taylor-series expansion of the relevant Westergaard stress function of the problem proved quite successful. Particular methods were developed for the determination of some coefficients of this function which play an important role in the evaluation of the stress field. Precautions were taken to avoid inherent errors due to the stress singularity and non-linear effects in the neighbourhood of the crack tip. The method was used for opening-mode as well as mixed-mode stress fields under static and dynamic loads. It is established that in spite of the inherent difficulties photoelasticity can successfully be used for analysing crack problems.

1. INTRODUCTION

Fracture mechanics has received much attention in the last two decades as a means of determining the useful life of structural components under various loading and environmental conditions. This discipline is based on the principle that all materials contain initial defects in the form of cracks, voids or inclusions which affect the load carrying capacity of engineering structures. Fracture in solids is initiated from such types of imperfection which cause high stress levels in that region. For a sufficiently high value of the local stress around the microcrack the atomic bonds at the crack edge may be broken. This leads to a growth of the flaw into a sizeable fracture surface which under certain favourable conditions can lead to complete failure of the solid. Due to the difficulties encountered in the analytical treatment of the strength of atomic bonds, an alternative, continuum mechanics, approach to the problem of crack propagation is usually adopted.

Since the failure process is the final stage of the deformed state of the body the quantitative assessment of crack propagation requires a complete knowledge of the local stress and strain fields in combination with a suitable fracture criterion. For the usual case of a two-dimensional plate with a through the thickness crack and subjected to in-plane loading, the crack tip stresses, due to the conditions of generalised plane stress or plane strain, can be expressed by a two-parameter set of equations. These parameters, called the stress intensity factors are functions of the crack dimensions and the applied loads. The critical values of the stress intensity factors are material constants and can be determined experimentally. Comparing the actual stress intensity factors with their critical values the conditions for unstable crack extension can be established.

Solution of the stress problem of a sharp crack within the framework of linear elasticity gives rise to infinite stresses at the singular crack tip where the radius of curvature is zero. In reality, the shape of the crack after deformation has a finite curvature at the tip and the stresses remain bounded. Thus, use of a large deformation theory would predict finite stresses at the crack tip. Furthermore, the existence of local plastic deformation reduces the stress concentration effect of the crack. For a small plastic zone in comparison with the crack length the stress distributions outside the plastic zone will not be seriously disturbed. This justifies the use of the singular solution for the description of the state of affairs near cracks.

The continuum theory of fracture assumes the existence of cracks in the solid and does not account for the conditions of crack initiation. Interest is then focused on the determination of the stress field with particular emphasis on the region near to the crack tip. This field is governed by the two stress intensity factors. The determination of these parameters is a problem of the mathematical theory of fracture and has been achieved for a variety of geometrical configurations and loading conditions of the cracked body. A survey of stress intensity factor solutions for a large number of two- and three-dimensional problems is provided in ref. 1.

Although analytical and/or numerical methods have been used extensively for determining stress intensity factors, experimental studies of fracture mechanics have been rather slow to develop. This is mainly due to the existing singularity at the crack tip which results in a rapid variation of the stress parameters in that region. Thus, with optical stress analysis methods, such as photoelasticity, moiré, interferometry, holography, etc., the steep variation of the stress field near the crack tip results in a steep variation of the fringes in the corresponding optical patterns. This makes the analysis and interpretation of the obtained patterns difficult. The direct application of these methods for the determination of stress intensity factors needs particular attention for the evaluation of the experimental results. At this point it should be mentioned that considerable progress towards the experimental solution of fracture mechanics problems has been achieved by the optical method of caustics developed by Theocaris.[2–4] In this method the stress singularity is transformed into an optical singularity, and creates a highly illuminated surface in space which when projected on a reference screen forms the caustic curve. The caustic is intimately related to the area very close to the crack tip and contains enough information for the determination of the state of affairs there.

The present work considers the problem of the photoelastic determination of the stress field around two- and three-dimensional cracks under static and dynamic loading. Particular methods and techniques for extracting stress intensity factors from photoelastic patterns are reviewed. The potential and limitations of photoelasticity for analysing fracture mechanics problems are analysed. Before proceeding to the analysis, a brief review of the preliminaries of fracture mechanics for a line crack in a two-dimensional body will be given.

2. THEORETICAL CONSIDERATIONS

The problem analysed first is that of a through the thickness crack in a two-dimensional body subjected to in-plane loading (Fig. 1). For this

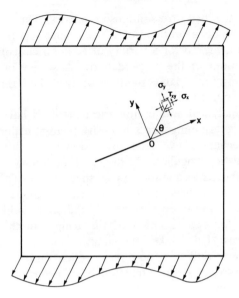

Fig. 1. Geometry and notation of a cracked plate.

case the stress field in the vicinity of the crack tip $\left(\dfrac{r}{a}\ll 1\right)$ can be described by the following equations:[5]

$$\sigma_x = \frac{1}{\sqrt{2\pi r}}\left[K_I\cos\frac{\theta}{2}\left(1-\sin\frac{\theta}{2}\sin\frac{3\theta}{2}\right)\right.$$
$$\left. - K_{II}\sin\frac{\theta}{2}\left(2+\cos\frac{\theta}{2}\cos\frac{3\theta}{2}\right)\right] - \sigma_{0x} \quad (1)$$

$$\sigma_y = \frac{1}{\sqrt{2\pi r}}\left[K_I\cos\frac{\theta}{2}\left(1+\sin\frac{\theta}{2}\sin\frac{3\theta}{2}\right)\right.$$
$$\left. + K_{II}\sin\frac{\theta}{2}\cos\frac{\theta}{2}\cos\frac{3\theta}{2}\right] \quad (2)$$

$$\tau_{xy} = \frac{1}{\sqrt{2\pi r}}\left[K_I\sin\frac{\theta}{2}\cos\frac{\theta}{2}\cos\frac{3\theta}{2}\right.$$
$$\left. + K_{II}\cos\frac{\theta}{2}\left(1-\sin\frac{\theta}{2}\sin\frac{3\theta}{2}\right)\right] \quad (3)$$

In these equations (r, θ) are the polar coordinates of the point considered with the origin defined at the crack tip, (K_I, K_{II}) are the opening-mode and sliding-mode stress intensity factors associated with symmetrical and non-symmetrical loading, respectively, of the plate with respect to the crack axis and σ_{0x}, the non-singular term in eqn (1) has been included to account for distant field stresses following Irwin's procedure.[6] Positive values of σ_{0x} tend to produce compressive stresses along the crack and negative values produce tensile stresses along the crack.

The maximum in-plane shear stress τ_m is related to the Cartesian components of stress by:

$$(2\tau_m)^2 = (\sigma_x - \sigma_y)^2 + 4\tau_{xy}^2 \tag{4}$$

Introducing the values of σ_x, σ_y, τ_{xy} from eqns (1)–(3) the following relation for τ_m is obtained:

$$(2\tau_m)^2 = \frac{1}{2\pi r}[(K_I \sin\theta + 2K_{II} \cos\theta)^2 + (K_{II} \sin\theta)^2]$$

$$+ \frac{2\sigma_{0x}}{\sqrt{2\pi r}} \sin\frac{\theta}{2}[K_I \sin\theta(1 + 2\cos\theta)$$

$$+ K_{II}(1 + 2\cos^2\theta + \cos\theta)] + \sigma_{0x}^2 \tag{5}$$

3. PHOTOELASTIC DETERMINATION OF K_I: TWO-PARAMETER METHODS

Post[7] and Wells and Post[8] were the first investigators to use photoelasticity for the determination of the state of stress and crack velocity in static and dynamic problems in the 1950s. Irwin[6] in a discussion of ref. 8 pointed out that for the photoelastic analysis of opening-mode crack tip stress fields two parameters at least are needed: the opening-mode stress intensity factor, K_I, and the normal stress, σ_{0x}, acting parallel to the direction of crack extension.

According to the photoelastic law the isochromatic fringe order N is related to the maximum in-plane shear stress, τ_m, by:

$$2\tau_m = \frac{Nf}{t} \tag{6}$$

where f is the stress-optical constant and t is the model thickness.

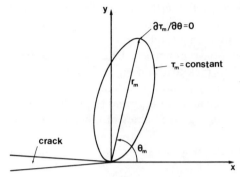

Fig. 2. Maximum distance r_m and angle of inclination θ_m of a crack tip isochromatic fringe loop.

From eqns (5) and (6) and for symmetrical loading ($K_{II} = 0$) we obtain:

$$(2\tau_m)^2 = \frac{K_I}{2\pi r}\sin^2\theta + \frac{2\sigma_{0x}K_I}{\sqrt{2\pi r}}\sin\theta\sin\frac{3\theta}{2} + \sigma_{0x}^2 \tag{7}$$

Irwin,[6] observing the geometry of the isochromatic loops around the crack tip, suggested measuring the distance from the crack tip to the farthest point on a given loop, r_m, together with its angle of inclination to the crack plane, θ_m (Fig. 2). The position of the farthest point on a given loop is dictated by:

$$\frac{\partial \tau_m}{\partial \theta} = 0 \tag{8}$$

which gives:

$$\sigma_{0x} = \frac{-K_I}{\sqrt{2\pi r_m}}\frac{\sin\theta_m\cos\theta_m}{\left(\cos\theta_m\sin\dfrac{3\theta_m}{2} + \frac{3}{2}\sin\theta_m\cos\dfrac{3\theta_m}{2}\right)} \tag{9}$$

From eqns (7) and (9) the two unknown quantities K_I and σ_{0x} are determined as:

$$K_I = \frac{Nf\sqrt{2\pi r_m}}{t\,\sin\theta_m}\left[1 + \left(\frac{2}{3\tan\theta_m}\right)^2\right]^{-1/2}\left(1 + \frac{2\tan\dfrac{3\theta_m}{2}}{3\tan\theta_m}\right) \tag{10}$$

$$\sigma_{0x} = -\frac{Nf}{t}\frac{\cos\theta_m}{\cos\left(\dfrac{3\theta_m}{2}\right)[\cos^2\theta_m + \frac{9}{4}\sin^2\theta_m]^{1/2}} \tag{11}$$

Note that for the determination of K_I and σ_{0x} from eqns (10) and (11) the quantities N, r_m and θ_m are needed. These quantities are measured from a single isochromatic loop. A typical isochromatic pattern in the vicinity of the crack tip is presented in Fig. 3. This figure was taken from a centre cracked plate of width, $b = 125$ mm, thickness, $t = 4\cdot9$ mm with a crack of length, $2a = 12\cdot5$ mm. The model was made of polycarbonate of bisphenol A (PCBA) and was subjected to uniaxial tension perpendicular to the crack axis.

Fig. 3. Isochromatic pattern in the close vicinity of the crack-tip for a thin plate with a central transverse crack submitted to uniaxial tension perpendicular to the crack axis.

In order to obtain an estimate of the influence of the error in measuring θ_m in the determination of K_I the non-dimensional quantity $K_I t / N f \sqrt{2\pi r_m}$ defined from eqn (10) was plotted in Fig. 4 against θ_m. The fringe tilt angle was varied in the interval $69 \cdot 4° \leqslant \theta_m \leqslant 148 \cdot 8°$. It is observed that the variation of K_I versus θ_m is very abrupt near the ends of the interval of θ_m, which means that a small error in measuring θ_m generates a large error in estimating K_I. Thus, the two-parameter method of Irwin predicts K_I with small error when θ_m lies in the previous interval, while for θ_m outside this interval the error increases rapidly and the method is not applicable. This error of K_I was shown to be of the order of $\pm 5\%$ for $73° \leqslant \theta_m < 139°$ and of the order of $\pm 2\%$ for $73 \cdot 5° < \theta_m < 134°$.[9]

In order to minimise errors of K_I when θ_m is close to either $69 \cdot 4°$ or $148 \cdot 8°$ Bradley and Kobayashi[10,11] used data obtained from two fringe loops. Introducing the quantity δ from relation:

$$\sigma_{0x} = \frac{\delta K_I}{\sqrt{\pi \alpha}} \tag{12}$$

they obtained for K_I:

$$K_I = \frac{f}{t} \frac{\sqrt{2\pi} \sqrt{r_1 r_2}(N_2 - N_1)}{f_2 \sqrt{r_1} + f_1 \sqrt{r_2}} \tag{13}$$

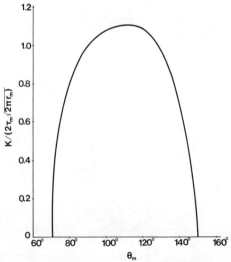

Fig. 4. Variation of the normalised stress intensity factor as a function of the tilt angle of the isochromatic pattern close to the crack tip.

where

$$f = [\sin^2 \theta + 2\delta \sqrt{2r/a} \sin \theta \sin (3\theta/2) + 2r\delta^2/a]^{1/2} \qquad (14)$$

Schroedl and Smith[12] used data from the isochromatic pattern along the line perpendicular to the crack plane ($\theta = 90°$). For this case eqn (7) reduces to:

$$(2\tau_m)^2 = \frac{K_I^2}{2\pi r} + \frac{K_I \sigma_{0x}}{\sqrt{\pi r}} + \sigma_{0x}^2 \qquad (15)$$

and solving for K_I we obtain:

$$K_I = \sqrt{\pi r} \left[\left[2\left(\frac{Nf}{t}\right)^2 - \sigma_{0x}^2 \right]^{1/2} - \sigma_{0x} \right] \qquad (16)$$

Furthermore, Schroedl and Smith simplified this equation by neglecting σ_{0x}^2 relative to $2(Nf/t)^2$ to obtain:

$$K_I = \sqrt{\pi r} \left[\sqrt{2}\left(\frac{Nf}{t}\right) - \sigma_{0x} \right] \qquad (17)$$

Applying this equation to two points (r_1, r_2) with fringe order (N_1, N_2) and solving for K_I we obtain:

$$K_I = \frac{f}{t} \sqrt{2\pi r_1} \frac{(N_1 - N_2)}{1 - (r_1/r_2)^{1/2}} \qquad (18)$$

If eqn (18) is applied to all pairs of fringe loops with specified values of N and r for each loop a set of values of K_I is determined. Then the average K_I is computed and the values of K_I which are outside a standard deviation limit are rejected and K_I is recomputed from the remaining values of K_I. For radii (r_1, r_2) measured without error the accuracy in predicting K_I by this method is of the order of $\pm 5\%$.[13]

Etheridge and Dally[9] in a critical review of the above three two-parameter methods indicated the limitations, the range of applicability and the errors in determining K_I by these methods. Thus, they concluded that all these methods can be used to determine K_I to within $\pm 5\%$ provided the tilt angle of the isochromatic loop is in the interval $73° \leq \theta_m < 139°$ and $r_m/a < 0.03$. For values of θ_m outside this range the two-parameter methods should not be employed. Furthermore, they indicated that the Bradley–Kobayashi differencing method gives the most accurate results when the radii of the fringe loops can be measured with better than 2% accuracy. When this condition is not satisfied the differencing method magnifies these errors and Irwin's method produces more accurate values of K_I. This latter method is exclusively employed when only one fringe loop is available for analysis. The differencing method of Schroedl and Smith employs a fixed value of θ_m and therefore the errors in θ_m measurement are eliminated.

4. FACTORS INFLUENCING DATA FOR K_I DETERMINATION

Before proceeding to more elaborate methods for determining K_I from photoelastic data an analysis of the factors influencing these data should be made. This would help the photoelastician to become more familiar with the problems inherent in selecting data from the photoelastic pattern and to use photoelasticity more effectively to determine K_I stress intensity factors.

In the previously described two-parameter methods the photoelastic determination of stress intensity factors was based on measurements on the isochromatic fringe pattern in the vicinity of the crack tip. The data should be taken from a zone dominated by the singular stresses. This data zone whose size varies widely from problem to problem depending upon the geometry must be very close to the crack tip. As we recede from the crack tip additional terms to those associated with the singular stresses are needed. On the other hand, if one measures too close to the crack tip invalid data may result. First of all in any analysis of crack problems a 'process' or 'core' region surrounding the crack tip must be introduced. This arises from the inability to describe in detail the state of affairs in the immediate vicinity of the crack tip. Indeed, the analytical solution of the stress and displacement field based on the theory of continuum mechanics must necessarily be defined at a finite distance away from the point where the influence of microstructure becomes dominant. Within the core region, the continuum mathematical model can no longer adequately describe the physical behaviour of the material, which is highly strained and may even be inhomogeneous. The radius of the core region for most engineering metal alloys was found to be of the order 10^{-3}–10^{-2} in. Outside the core region, the stress field can be adequately described by the continuum mechanics solution. Hence, the boundary of the core zone serves to separate the outside material assumed to behave elastically from the inside material whose mechanical properties are unknown. The size of the core region reflects the material property.

The zone of valid data should, therefore, extend outside the core region. Also, additional factors inherent in the use of the photoelastic method for solving crack problems put further restrictions in the determination of the data zone. Thus, the isochromatic pattern is influenced by the following factors which alter its meaning in the crack tip area:

4.1. Notch-end Geometry
In photoelastic work an artificial notch is usually made in the specimen to simulate the crack. This change in the geometry obviously alters the stress field in the vicinity of the notch end. Also, when natural cracks are used they open under tension into a geometry approximating a narrow ellipse. A thorough analysis of the influence of the notch end geometry upon the photoelastic fringe pattern ahead of the notch end was made by Schroedl *et al.*[13] Using the Kolosoff–Inglis solution for an elliptical hole in a plate subjected to biaxial tension they computed the maximum in-plane shear stress, τ_m, for various root radii of the ellipse and compared this stress with that obtained for a crack with length equal to the major axis of the ellipse. They obtained the error in τ_m for the line passing through the notch end and perpendicular to the notch axis for a notch root radius $\rho = 0\cdot01$ as 2% for $r/a = 0\cdot04$, but this increases to 7% for $r/a = 0\cdot02$. They came to the conclusion that the fringe pattern is significantly influenced by the curvature of the notch inside a distance of one root radius ahead of the notch end. This result was previously obtained by Liebowitz.[14] Figure 5 presents the effect of finite root radius on the fringe pattern.

Furthermore, in another work Schroedl and Smith[15] studied various notch geometries (rectangular, rhombus, blunted triangular) and compared the values of the stress intensity factors obtained for various distances ahead of the notch end with those of a crack of the same overall length. They came to the conclusion that when possible, natural flaws should be used to simulate the line crack. In such a case data obtained outside of $r/a \simeq 10\rho/a$ will be free of notch effects. For the case of rectangular notches data can be taken outside a zone resulting from the elliptical notch with root radius equal to the half width of the rectangular notch. Analogous observations are valid for the blunted triangular notch. Generally, the stress intensity factor determined from

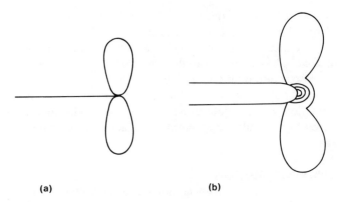

(a)   (b)

Fig. 5. Effect of the crack tip radius on the fringe pattern.

sharp angular notches is underestimated unless data are taken extremely close in.

4.2. Crack Tip Singularity
An enhanced density of isochromatic fringes is created in the area close to the crack tip due to the high stress concentration there. Precautions should be taken that the order of isochromatics in the data zone does not exceed the linear limit in the stress-fringe curve of the model material.

4.3. Triaxiality of Stress Field
For a cracked plate of finite thickness the state of stress and deformation is nearly plane strain near the crack tip, while it is characterised by generalised plane stress remote from the crack tip. Between these areas a transition zone exists in which the stress state is three-dimensional. Brown and Srawley[16,17] and Irwin[18] have indicated that the stress intensity factor obtained photoelastically from data in the transition zone must be multiplied by $\sqrt{1-\nu^2}$ to obtain the two-dimensional stress intensity factor for a thin plate with ν the value of the Poisson's ratio. When the stress-freezing technique is used $\nu = 0\cdot5$ and if we want to obtain the stress intensity factor for $\nu = 0\cdot3$ we may use the inverse of the above conversion. Thus a complete conversion factor $\sqrt{1-0\cdot5^2}/\sqrt{1-0\cdot3^2} = 0\cdot91$ is obtained to correct the test result due to the mismatch of Poisson's ratio in stress freezing work and the structural material.

5. MULTIPLE-PARAMETER METHODS FOR K_I DETERMINATION

From the above discussion it is seen that measurements in the close vicinity of the crack tip are prone to the deleterious influence of various factors, such as the radius of the root at the crack tip, the existing high stress variation, the rapid change of refractive index, material imperfections, stresses which may exceed the limit of linearity of the stress-optic law or high stresses due to impending plasticity, triaxiality of the stress field and finally eventual inhomogeneities and anisotropy of the material. All these factors alter the physical meaning of isochromatics which differ from those corresponding to the respective purely elastic generalised plane stress problem. It is therefore attractive to gather data far enough away from the crack tip where the influence of all these factors on the isochromatic pattern becomes negligible. This requires additional terms for the description of the stress field than those associated with the singular stress components.

Theocaris and Gdoutos[19] suggested the use of a Taylor-series expansion of the stress components for a more accurate description of the isochromatic pattern far away from the crack tip. They established a quantitative relation between the exact solution of the stress field in the vicinity of the crack tip derived from Westergaard's formulation and the singular solution and the results obtained were correlated with photoelastic data to determine stress intensity factors. The maximum in-plane shear stress for the exact and singular solution were calculated and compared for a cracked plate subjected to uniaxial and biaxial tension. These results established the region where measurements in the isochromatic pattern are possible for the accurate determination of the stress intensity factor. Furthermore, extrapolation laws for the analysis of the region near the crack tip from data obtained from the far-field of isochromatics were obtained.

Smith and his coworkers[20–22] put the maximum in-plane shear stress, τ_m, along $\theta = 90°$ in a Taylor-series expansion

$$\tau_m = \frac{A}{r^{1/2}} + B + \sum_{N=1}^{M} C_N r^N \tag{19}$$

and developed a computer program to determine the coefficients A, B, C_N. The program evaluates first A, then A, B, then A, B, C_1, etc., by a least-squares analysis of the fringe data until the series in truncated. By applying this method to a variety of two- and three-dimensional problems no specific truncation criterion has been found to be applicable. The lowest order curve that best fits the experimental data was used. Usually, three terms of the series in eqn (19) were adequate for the analysis. Use of more than three terms was justified in problems where a crack approached a free boundary.

Dally and his associates[23,24] developed a three- and four-parameter method to determine the stress intensity factor, K_I. The four parameters are: the stress intensity factor, K_I, a normal stress, $\sigma_{0x} = \alpha K_I/\sqrt{2\pi a}$, parallel to the direction of crack extension, a fictitious crack length, a, and a parameter β added to the stress function in order to model the effect of near field boundaries and boundary loading. This makes possible the analysis of fringe loops further away from the crack tip. Thus, the Westergaard stress function, $Z(z)$, is put in the form:

$$Z(z) = \frac{K_I}{\sqrt{2\pi z}} [1 + \beta(z/a)] \tag{20}$$

$$z = re^{i\theta}$$

From $Z(z)$ the Cartesian stress components are given by:

$$\sigma_x = \text{Re } Z - y \text{ Im } Z' \tag{21}$$

$$\sigma_y = \text{Re } Z + y \text{ Im } Z' \tag{22}$$

$$\tau_{xy} = -y \text{ Re } Z' \tag{23}$$

where Re and Im represent real and imaginary parts and $Z' = dZ/dz$.

Substituting eqn (20) into eqns (21), (22) and (23) and adding the uniform stress σ_{0x} to σ_x gives:

$$\sigma_x = \frac{K_I}{\sqrt{2\pi r}} \left[\cos\frac{\theta}{2} \left(1 - \sin\frac{\theta}{2} \sin\frac{3\theta}{2} \right) \right.$$
$$\left. + \cos\frac{\theta}{2} \left(1 + \sin^2\frac{\theta}{2} \right) \beta(r/a) + \alpha\sqrt{r/a} \right] \tag{24}$$

$$\sigma_y = \frac{K_I}{\sqrt{2\pi r}} \left[\cos\frac{\theta}{2} \left(1 + \sin\frac{\theta}{2} \sin\frac{3\theta}{2} \right) \right.$$
$$\left. + \cos\frac{\theta}{2} \left(1 - \sin^2\frac{\theta}{2} \right) \beta(r/a) \right] \tag{25}$$

$$\tau_{xy} = \frac{K_I}{\sqrt{2\pi r}} \sin\frac{\theta}{2} \cos\frac{\theta}{2} \left[\cos\frac{3\theta}{2} - \beta(r/a) \cos\frac{\theta}{2} \right] \tag{26}$$

The maximum in-plane shear stress given from eqn (4) is expressed by:

$$\tau_m^2 = \frac{K_I^2}{8\pi r} \left[\sin^2 \theta [1 - 2\beta(r/a)\cos\theta + \beta^2(r/a)^2] - 2\alpha\sqrt{r/a}\sin\theta \right.$$

$$\left. \times \left[\sin\frac{3\theta}{2} - \beta(r/a)\sin\frac{\theta}{2} \right] + \alpha^2(r/a) \right] \tag{27}$$

Introducing the stress-optical law of eqn (6) the position r of the isochromatic fringe of order N is:

$$r = \frac{\gamma^2}{2\pi} \left[\sin^2 \theta [1 - 2\beta(r/a)\cos\theta + \beta^2(r/a)^2] - 2\alpha\sqrt{r/a}\sin\theta \right.$$

$$\left. \times \left[\sin\frac{3\theta}{2} - \beta(r/a)\sin\frac{\theta}{2} \right] + \alpha^2(r/a) \right] \tag{28}$$

where

$$\gamma = \frac{K_I t}{Nf} \tag{29}$$

Equation (28) can be solved in a computer to give r as a function of θ with γ, α, a and β as parameters. Physical solutions of eqn (28) should give real values of r such that $r/a < 1$, so that the Westergaard approach is valid. The four parameters of the method can be adjusted so that the analytical isochromatics match the experimental ones. When a close fit is achieved the stress intensity factor K_I is determined. This analysis permits the fringe loops to tilt, stretch and become unsymmetrical, so that they can be used to determine K_I for a wide variety of specimen geometries and loading conditions.

This method was used to obtain static and dynamic values of K_I in Homalite 100 with great success.

6. MIXED-MODE STRESS INTENSITY FACTORS

For the general case of an arbitrarily oriented crack in a two-dimensional plate subjected to in-plane loads the stress field in the vicinity of the crack tip is governed by both the opening-mode, K_I, and the sliding mode, K_{II}, stress intensity factors. Knowledge of these quantities enables the determination of the crack growth pattern and the critical load for crack initiation through a suitable fracture criterion. The author[25] solved a variety of crack problems of engineering importance in which mixed-mode conditions prevail and determined the fracture characteristic quantities using the strain energy density criterion. Because of analytical difficulties the photoelastic determination of K_I, K_{II} is of great importance.

Smith and Smith[26] were the first to use photoelasticity to determine K_I and K_{II}. Introducing the values of the stress components σ_x, σ_y, τ_{xy} from eqns (1)–(3) into eqn (4) the maximum in-plane shear stress τ_m is given by:

$$\tau_m = \frac{1}{2\sqrt{2\pi r}} [\sin^2 \theta K_I^2 + 2 \sin 2\theta K_I K_{II} + (4 - 3 \sin^2 \theta) K_{II}^2]^{1/2} \quad (30)$$

In deriving eqn (30) the non-singular term σ_{0x} was not taken into account. Figure 6 presents a photograph of the isochromatics in the vicinity of the crack tip for a thin plate with an edge crack of inclination $\beta = 50°$ to the applied tensile stress. If eqn (30) is now applied to two points of the isochromatic fringe pattern with known values of τ_m ($\tau_m = Nf/2t$), K_I and K_{II} can be derived from the solution of the resulting system of two equations.

As in the case of the opening-mode deformation the distance of the most remote point on a given isochromatic loop from the crack tip is determined from eqn (8). Differentiating τ_m from eqn (30) with respect to θ and equating the derivative to zero we obtain:

$$\left(\frac{K_{II}}{K_I}\right)^2 - \frac{4}{3}\left(\frac{K_{II}}{K_I}\right)\cot 2\theta_m - \frac{1}{3} = 0 \quad (31)$$

Equation (31) defines a relationship between the angle θ_m of the axis of symmetry of the fringe loops at the crack tip and the ratio K_{II}/K_I. Thus, measurement of θ_m from a photoelastic experiment enables the determination of K_{II}/K_I. The variation of K_{II}/K_I versus θ_m is presented in Fig. 7.

Both eqns (30) and (31), however, are valid very close to the crack tip in an area where measurements on the isochromatic pattern are prone to errors due to the factors analysed previously. This induced Gdoutos and Theocaris[27] to develop a new evaluation method of the isochromatic fringe pattern which bypasses the measurements on the isochromatic pattern near to the crack tip and uses data from the far region. Suitable extrapolation laws are established by comparing the near and the far to the crack tip isochromatic patterns. Then the singular solution is used to determine K_I and K_{II}.

Fig. 6. Isochromatic pattern for a thin plate with an edge crack of inclination 50° to the applied uniaxial uniform stress.

Dally and Sanford[28,29] introduced the far-field stress σ_{0x} into the photoelastic determination of K_I and K_II. For this case eqn (30) takes the form:

$$\tau_m^2 = \frac{1}{8\pi r} [\sin^2 \theta K_\mathrm{I}^2 + 2 \sin 2\theta K_\mathrm{I} K_\mathrm{II} + (4 - 3 \sin^2 \theta) K_\mathrm{II}^2]$$

$$+ \frac{\sigma_{0x}}{2\sqrt{2\pi r}} \sin \frac{\theta}{2} [\sin \theta (1 + 2 \cos \theta) K_\mathrm{I} + (1 + 2 \cos^2 \theta + \cos \theta) K_\mathrm{II}]$$

$$+ \frac{\sigma_{0x}^2}{4} \tag{32}$$

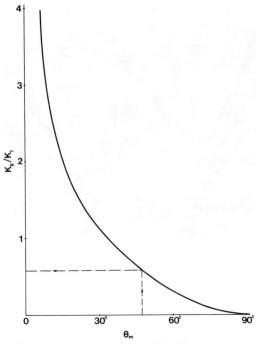

Fig. 7. Variation of the ratio K_{II}/K_I of the stress intensity factors versus the tilt angle θ_m.

Based on this equation they have established the effect of K_I, K_{II} and σ_{0x} on the shape and size of the fringe loops. It was shown that the effects are so pronounced that the presence and the sense of either K_{II} or σ_{0x} can be indicated by a rough examination of the fringe pattern. Six forms of isochromatic patterns associated with K_I alone, K_{II} alone, K_I and σ_{0x}, K_I and K_{II}, K_{II} and σ_{0x} and K_I, K_{II} and σ_{0x} were examined and their characteristic features were identified. Equation (6) was then used to determine K_I, K_{II} and σ_{0x} from the isochromatic patterns. For the case that all these three parameters are present eqn (32) is non-linear in terms of K_I, K_{II} and σ_{0x}. The Newton–Raphson numerical method was used to solve this equation. This method was further used by Sanford.[30]

7. DYNAMIC ANALYSIS

Earlier investigations[31,32] into the use of photoelasticity to determine stress intensity factors and crack velocities of propagating cracks were based on the static solution of the stress field around the crack tip. Later on dynamic stresses were introduced into the analysis. The dynamic stress components for a crack propagating at a constant velocity c are given by[33]

$$\sigma_x = \frac{1+r_2^2}{4r_1r_2-(1+r_2^2)^2} \frac{K}{\sqrt{2\pi r}} \left[(2r_1^2 - r_2^2 + 1)f_1(\theta)\cos(\theta_1/2) \right.$$
$$\left. - \frac{4r_1r_2}{1+r_2^2} f_2(\theta)\cos(\theta_2/2) \right] + \sigma_{0x} \tag{33}$$

$$\sigma_y = \frac{1+r_2^2}{4r_1r_2-(1+r_2^2)^2} \frac{K}{\sqrt{2\pi r}} \left[-(1+r_2^2)f_1(\theta)\cos(\theta_1/2) \right.$$
$$\left. + \frac{4r_1r_2}{1+r_2^2} f_2(\theta)\cos(\theta_2/2) \right] \tag{34}$$

$$\tau_{xy} = \frac{(1+r_2^2)2r_1}{4r_1r_2-(1+r_2^2)^2} \frac{K}{\sqrt{2\pi r}} [f_1(\theta)\sin(\theta_1/2) - f_2(\theta)\sin(\theta_2/2)] \tag{35}$$

where:
$$f_1(\theta) = [1 - (\dot{a}/c_1)^2 \sin^2\theta]^{-1/4}$$
$$f_2(\theta) = [1 - (\dot{a}/c_2^2) \sin^2\theta]^{-1/4}$$
$$r_1^2 = 1 - (\dot{a}/c_1)^2 \tag{36}$$
$$r_2^2 = 1 - (\dot{a}/c_2)^2$$
$$\tan\theta_1 = r_1 \tan\theta$$
$$\tan\theta_2 = r_2 \tan\theta$$

and c_1 and c_2 are the dilatational and shear-wave velocities, respectively.

Introducing the stresses σ_x, σ_y, τ_{xy} into the photoelastic law expressed from eqn (6) the following equation, which describes the shape of the dynamic isochromatic pattern, is obtained:

$$\left(\frac{Nf\sqrt{2\pi r}}{2tB_1K} \right)^2 - \left(B_2 + \frac{\sigma_{0x}\sqrt{2\pi r}}{2B_1K} \right)^2 - B_3^2 = 0 \tag{37}$$

where B_1, B_2 and B_3 are functions of θ and the velocity ratios.

$$B_1 = \frac{1+r_2^2}{4r_1r_2-(1+r_2^2)^2}$$

$$B_2 = (1+r_1^2)f_1(\theta)\cos(\theta_1/2) - \frac{4r_1r_2}{1+r_2^2} f_2(\theta)\cos(\theta_2/2) \tag{38}$$

$$B_3 = 2r_1[f_1(\theta)\sin(\theta_1/2) - f_2(\theta)\sin(\theta_2/2)]$$

Kobayashi and Mall[34] estimated the errors introduced in the determination of the dynamic stress intensity factor when the near to the crack tip static stress field was used. They concluded that the error introduced is small for crack propagation velocities, c, less than $c/c_1 = 0 \cdot 15$. Investigations[35-39] by Dally, Etheridge and Rossmanith showed that significant differences in K occur for higher crack velocities and the above analysis was used to determine the dynamic stress intensity factor. Gdoutos[40] pointed out that for duplex specimens studied in ref. 39 a value of the crack tip stress singularity different from $-0 \cdot 5$ should be introduced in the analysis when the crack meets the interface.

Extensive studies were conducted at the Photomechanics Laboratory of the University of Maryland to determine the stress intensity factor, K, and the crack propagation velocity, $\dot{a}$. The main objective of these studies was directed towards establishing a relation between K and $\dot{a}$. The dynamic photoelastic-fringe patterns were recorded by a Cranz–Schardin type high-speed camera. Figure 8 presents the general form of the $K-\dot{a}$ curve. These curves were determined for six different fracture specimen geometries in order to establish the influence of the specimen type on the $K-\dot{a}$ relationship. It was found that for crack velocities below $300 \, \mathrm{m \, s^{-1}}$ a relationship $K-\dot{a}$ independent of the specimen geometry exists for Homalite 100.

In the previous analysis the stress field around the propagating crack

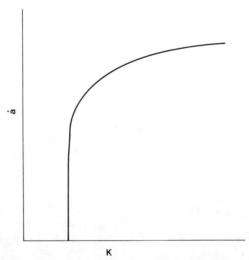

Fig. 8. Sketch of a typical $K-\dot{a}$ curve for Homalite 100.

is purely elastic. Betser *et al.*[41] determined the effect of small-scale crack tip yielding on the dynamic stress intensity factor using a Dugdale-strip yield zone in the propagating crack. They found that dynamic isochromatics for small-scale yielding result in a lower apparent dynamic stress intensity factor and obtained plasticity correction factors for *K*.

The influence of the K_{II} stress intensity factor on the photoelastic pattern surrounding a propagating crack was studied by Kobayashi and Ramulu.[42] This case occurs when the running crack exhibits moderate asymmetry due to slightly curved crack propagation paths, caused by the coexistence of a small K_{II} with K_I. Using the mixed-mode dynamic solution of the stress field a relationship between K_I, K_{II} and σ_{0x} was obtained in the same way as for the static case. Based on this relation and a least-squares method the values of K_I and K_{II} were determined and the errors introduced in using a mixed-mode static crack tip solution were estimated.

8. THREE-DIMENSIONAL CRACKS

The photoelastic determination of stress intensity factors in three-dimensional crack problems is of particular importance due to the difficulties encountered in the analytical or numerical solutions. With the development of high-speed digital computers several approximate methods have been developed for such problems. These methods include finite elements, finite differences, integral-equation methods, boundary integrals and hybrid combinations.

In three dimensions the crack front is generally curved and the stress intensity factor varies along the crack border. Kassir and Sih[43] have shown that the singular elastic stress field around the crack border can be expressed in the same way as in the two-dimensional case when a local moving rectangular Cartesian coordinate system with its axes defined by the tangent, the normal and the binormal of the crack border at the point considered is introduced.

The frozen stress photoelastic technique combined with the previously developed methods for two-dimensional crack problems has been used extensively by Smith and his coworkers[44–48] to analyse cracks in three dimensions. In a large number of recent publications they studied many problems of practical interest and demonstrated the feasibility of photoelasticity in solving three-dimensional crack problems. Among these problems are the semi-elliptical crack in a plate subjected to cylindrical bending, a part-circular crack in a plate under remote tension, a hole crack, a pressure-vessel corner nozzle and a nozzle in a flat plate, amongst others. Using the Taylor series method of expanding the maximum shear stress τ_m along the line $\theta = 90°$ they obtained the stress intensity factor along the crack border. For more details on the technique, the models, the loading devices and the special characteristic features encountered in the solution of the above problems the interested reader is referred to refs 44–48.

REFERENCES

1. Sih, G. C., *Handbook of Stress-Intensity Factors*, Institute of Fracture and Solid Mechanics, Lehigh University, Bethlehem, PA, 1973.
2. Theocaris, P. S., Local yielding around a crack-tip in Plexiglas, *J. Appl. Mech., Trans. ASME Series E*, **37** (1970), 409–15.
3. Theocaris, P. S. and Gdoutos, E. E., An optical method for determining opening-mode and edge sliding-mode stress-intensity factors, *J. Appl. Mech. Trans. ASME Series E*, **39** (1972) 91–7.
4. Theocaris, P. S., The method of caustics applied to the solution of engineering problems, In: *Recent Developments in Stress Analysis*, ed. G. S. Holister, Applied Science Publishers, London, 1979, pp. 27–63.
5. Paris, P. C. and Sih, G. C., Stress analysis of cracks, *ASTM STP 381* (1964), pp. 30–81.
6. Irwin, G. R., Discussion of the paper: The Dynamic Stress Distribution Surrounding a Running Crack—A Photoelastic Analysis, *Proc. Soc. Exp. Stress Analysis*, **16** (1958), 93–6.
7. Post, D., Photoelastic stress analysis for an edge crack in a tensile field, *Proc. Soc. Exp. Stress Analysis*, **12** (1954), 99–116.
8. Wells, A. A. and Post, D., The dynamic stress distribution surrounding a running crack—A photoelastic analysis, *Proc. Soc. Exp. Stress Analysis*, **16** (1958), 69–92.
9. Etheridge, J. M. and Dally, J. W., A critical review of methods for determining stress-intensity factors from isochromatic fringes, *Exp. Mech.*, **17** (1977), 248–54.
10. Bradley, W. B., A photoelastic investigation of dynamic brittle fracture, PhD Thesis, University of Washington, 1969.
11. Bradley, W. B. and Kobayashi, A. S., An investigation of propagating cracks by dynamic photoelasticity, *Exp. Mech.*, **10** (1970), 106–13.
12. Schroedl, M. A. and Smith, C. W., Local stress near deep surface flaws under cylindrical bending fields, *ASTM STP 536*, (1973), pp. 45–63.
13. Schroedl, M. A., McGowan, J. J. and Smith, C. W., An assessment of factors influencing data obtained by the photoelastic stress freezing technique for stress fields near crack tips, *J. Engng. Fract. Mech.*, **4** (1972), 801–9.

14. Liebowitz, H., Vanderveldt, H. and Sanford, R. J., Stress concentrations due to sharp notches, *Exp. Mech.*, **7** (1967), 513–17.
15. Schroedl, M. A. and Smith, C. W., A study of near and far field effects in photoelastic stress intensity determination, *J. Engng. Fract. Mech.*, **7** (1975), 341–55.
16. Brown, W. F., Jr. and Srawley, J. E., Fracture toughness testing, *ASTM STP 381*, (1964), pp. 133–96.
17. Brown, W. F., Jr., Plane strain crack toughness testing of high strength metallic materials, *ASTM STP 410* (1967), pp. 1–65.
18. Irwin, G. R., Measurement challenges in fracture mechanics, *William Murray Lecture, SESA Fall Meeting*, Indianapolis, 1973.
19. Theocaris, P. S. and Gdoutos, E. E., A photoelastic determination of K_I stress intensity factors, *Engng Fract. Mech.*, **7** (1975), 331–9.
20. Schroedl, M. A., McGowan, J. J. and Smith, C. W., Determination of stress-intensity factors from photoelastic data with applications to surface-flaw problems, *Exp. Mech.*, **14** (1974), 392–9.
21. Smith, C. W., Use of three-dimensional photoelasticity in fracture mechanics, *Exp. Mech.*, **13** (1973), 539–44.
22. Schroedl, M. A., McGowan, J. J. and Smith, C. W., Use of a Taylor series correction method in photoelastic stress intensity determinations, *Spring Meeting SESA*, Detroit, 1974.
23. Etheridge, J. M. and Dally, J. W., A three-parameter method for determining stress intensity factors from isochromatic fringe patterns, *J. Strain Anal.*, **13** (1978), 91–4.
24. Etheridge, J. M., Dally, J. W. and Kobayashi, T., A new method of determining the stress intensity factor K from isochromatic fringe loops, *Engng Fract. Mech.*, **10** (1978), 81–93.
25. Gdoutos, E. E., *Problems of Mixed-Mode Crack Propagation*, Martinus Nijhoff, The Hague, 1984, pp. i–xiv, 1–2·4.
26. Smith, D. G. and Smith, C. W., Photoelastic determination of mixed mode stress intensity factors, *Engng Fract. Mech.*, **4** (1972), 357–66.
27. Gdoutos, E. E. and Theocaris, P. S., A photoelastic determination of mixed-mode stress-intensity factors, *Exp. Mech.*, **18** (1978), 87–96.
28. Dally, J. W. and Sanford, R. J., Classification of stress-intensity factors from isochromatic-fringe patterns, *Exp. Mech.*, **18** (1978), 441–8.
29. Sanford, R. J. and Dally, J. W., A general method for determining mixed-mode stress intensity factors from isochromatic fringe patterns, *Engng Fract. Mech.*, **11** (1979), 621–33.
30. Sanford, R. J., Application of the least-squares method to photoelastic analysis, *Exp. Mech.*, **20** (1980), 192–7.
31. Kobayashi, A. S. and Chan, C. F., A dynamic photoelastic analysis of dynamic-tear-test specimen, *Exp. Mech.*, **16** (1976), 176–81.
32. Kobayashi, A. S., Mall, S. and Lee, M. H., Fracture dynamics of wedge-loaded DCB specimen, cracks and fracture, *ASTM STP 601* (1976), pp. 274–90.
33. Irwin, G. R., Constant speed semi-infinite tensile crack opened by a line force P a distance b from the leading edge of the crack tip, Lehigh University Lecture Notes, 1968.

34. Kobayashi, A. S. and Mall, S., Dynamic fracture toughness of Homalite 100, *Exp. Mech.*, **18** (1978), 11–18.
35. Etheridge, J. M., Determination of the stress intensity factor *K* from isochromatic fringe loops, PhD Thesis, University of Maryland, 1976.
36. Rossmanith, H. P. and Irwin, G. R., Analysis of dynamic isochromatic crack-tip stress patterns, University of Maryland Report, 1979.
37. Irwin, G. R., Dally, J. W., Kobayashi, T., Fourney, W. L., Etheridge, M. J. and Rossmanith, H. P., On the determination of the $\dot{a}-K$ relationship for birefringent polymers, *Exp. Mech.*, **19** (1979), 121–8.
38. Dally, J. W., Dynamic photoelastic studies of fracture, *Exp. Mech.*, **19** (1979), 349–61.
39. Dally, J. W. and Kobayashi, T., Crack arrest in duplex specimens, *Int. J. Solids Struct.*, **14** (1979), 121–9.
40. Gdoutos, E. E., Determination of stress intensity factors during crack arrest in duplex specimens, *Int. J. Solids Struct.*, **17** (1981), 683–5.
41. Betser, A. A., Kobayashi, A. S., Lee, O. S. and Kang, B. S.-J., Crack-tip dynamic isochromatics in the presence of small-scale yielding, *Exp. Mech.*, **22** (1982), 132–8.
42. Kobayashi, A. S. and Ramulu, M., Dynamic stress-intensity factors for unsymmetric dynamic isochromatics, *Exp. Mech.*, **21** (1981), 41–8.
43. Kassir, M. and Sih, G. C., Three-dimensional stress distribution around an elliptical crack under arbitrary loadings, *J. Appl. Mech., Trans ASME Series E*, **33** (1966), 601–11.
44. Smith, C. W., Use of three-dimensional photoelasticity in fracture mechanics, *Exp. Mech.*, **13** (1973), 539–44.
45. Smith, C. W., Jolles, M. and Peters, W. H., Stress intensities for nozzle cracks in reactor vessels, *Exp. Mech.*, **17** (1977), 449–54.
46. Smith, C. W., McGowan, J. J. and Peters, W. H., A study of crack-tip nonlinearities in frozen-stress fields, *Exp. Mech.*, **18** (1978), 309–15.
47. Smith, C. W., Stress intensity and flaw-shape variations in surface flaws, *Exp. Mech.*, **20** (1980), 126–33.
48. Smith, C. W., Photoelasticity in fracture mechanics, *Exp. Mech.*, **20** (1980), 390–6.

7

CONVENTIONAL THREE-DIMENSIONAL PHOTOELASTICITY: A REVIEW OF PRINCIPLES AND MATERIALS

S. A. Paipetis

Department of Mechanical Engineering, University of Patras, Greece

ABSTRACT

The principles, techniques and materials used in conventional three-dimensional photoelasticity based on stress-freezing and slicing are reviewed. A quantitative description of the stress-freezing phenomenon is given on the basis of the linear viscoelastic behaviour of the respective materials, along with phenomena associated with slicing. The various resin systems are examined in order to determine the ones best conforming with the requirements of three-dimensional photoelastic applications. Machining operations on such systems are described and evaluated. Particular problems, such as polymerisation exotherm and shrinkage, as well as cracking phenomena, are examined, along with the cost-effectiveness of three-dimensional photoelasticity as a stress-analysis tool.

1. INTRODUCTION

By 'conventional three-dimensional photoelasticity' we mean the well-known stress-freezing method in which a photoelastic material is heated until it reaches its rubbery state, and is then loaded and cooled down to the ambient temperature at a slow rate. The load is then removed, but the stresses developed remain 'locked-in' or 'frozen' and the specimen can be sliced while the corresponding photoelastic field remains unaffected. The photoelastic analysis can then proceed by means of the examination of the photoelastic fields of the individual slices.

The method, the principles of which began to develop more than 50 years ago,[1-4] turned out to be a most powerful tool for the experimental solution of three-dimensional problems of elasticity. However, it is time-consuming and consequently expensive, and the development of computer-supported methods, such as the finite-element method appeared to provide quick and accurate solutions for a large number of structural problems. Three-dimensional photoelasticity, however, still retains its value both in research and in engineering design, under conditions which render it cost-effective.[6]

As stated, the method is based on the phenomenon of stress freezing for which there are many misconceptions as far as explanations of the phenomenon are concerned. In fact, what is considered as a 'locked-in' or 'frozen' stress field is only the initial stage of a very slow creep recovery procedure, which would take tens or even hundreds of years to be completed. Since this procedure can hardly be affected by the application of any short-duration external loads, the specimen can be sliced or even machined without any influence on the residual photoelastic pattern, provided that no temperature rise occurs. The viscoelastic nature of photoelastic polymers lies at the heart of stress-freezing phenomena, a fact which appears to be ignored, even by eminent photoelasticians.

Using the empirical explanation of the stress-freezing phenomenon given by Kuske,[2] the behaviour of the photoelastic material is simulated by that of a wax-impregnated sponge. This material deforms when loaded at a temperature higher than the melting point of wax and remains deformed when it cools down under load with the wax solidifying again. It is interesting to note that the dual nature of photoelastic polymers is reflected in this model.

The fact that the stress-freezing technique is in reality based on a creep procedure, greatly accelerated by the increased temperature, was verified by Durelli and Lake,[7] who introduced 'a new creeping method' for three-dimensional photoelasticity, i.e. they let the material creep under load for a considerable length of time and noticed that slicing did not affect the photoelastic fields in the specimen. Naturally, creep recovery can be expected to occur in a rather short time, something avoided with the conventional method.

Since Poisson's ratio for photoelastic polymers in their rubbery state tends to 0·5, no displacement measuring methods can be applied in combination with the photoelastic data, while the effect of Poisson's ratio cannot in general be neglected in the presence of high stress gradients. Therefore, attempts have been made to obtain stress-freezing at temperatures lower than the glass transition temperature of the material,[8] a method suffering the same drawbacks as the previous one.

Concerning Poisson's ratio, it is interesting to note a remark, in a work by Frocht and Guernsey,[9] that 'unfortunately, the values of ν for the photoelastic materials available in this country are approximately equal to 0·5 at the stress-freezing temperature and the method (i.e. of measuring displacements in annealed specimen) breaks down'.[4]

In the following a review is given of the various steps of the stress-freezing process.

2. QUANTITATIVE DESCRIPTION OF THE STRESS-FREEZING PHENOMENON

Consider a polymeric birefringent material, exhibiting linear visco-elastic behaviour and subjected to uniaxial creep. The stress–strain relation will be:

$$\varepsilon(t) = \sigma_0 J(t) \tag{1}$$

where $\varepsilon(t)$ is time-dependent strain, σ_0 is constant stress and $J(t)$ is creep-compliance. The latter can be obtained as a *composite* or *master curve* from short-time measurements at different temperatures by the method of reduced variables.[10] An example of master curves for a class of hot-setting epoxy resins is presented in Fig. 1. With temperatures varying from ambient to about 150°C, the simple forms displayed are produced. The flat initial and final parts of the curves correspond to the glassy and rubbery states of the materials, exhibiting elastic and rubber-elastic behaviour, respectively, while the middle part is the transition region, with the point of steepest slope corresponding to the *glass-transition* or *β-transition* temperature T_g. In the latter region, the materials exhibit strongly viscoelastic behaviour.

Similar curves can be constructed in order to describe the time-dependent birefringent properties of the material; for example, the strain-optical relationship corresponding to eqn (1) is:

$$R(t) = \sigma_0 \, . \, C_\varepsilon(t) \tag{2}$$

where $R(t)$ is the birefringence and $C_\varepsilon(t)$ is the strain-optical coefficient, exhibiting the same time-dependence as creep compliance, $J(t)$.

Now, let the material creep, following the law of eqn (1), until it reaches its rubbery state in time t_r and then let the load be removed, i.e. let a load $-\sigma_0$ be added. According to Boltzmann's superposition principle[10]

$$\varepsilon(t) = \sigma_0 \cdot J(t) + \sum_{i=1}^{n} \Delta\sigma_i J(t - t_i) \qquad (3)$$

the material will start recovering, according to the equation

$$\varepsilon(t) = \sigma_0[J(t) - J(t - t_r)] \qquad (4)$$

suggesting that time t_r is needed for the material to fully recover. Since this time is usually very long, of the order of tens or even hundreds of years, the recovery procedure is very slow, and the respective stress/strain fields along with the birefringence appear to be 'locked-in' or 'frozen' in the material.

If, subsequently, at time $t_p > t_r$, a very high load σ_p of very short duration Δt_p be applied, eqn (4) assumes the form:

$$\varepsilon(t) = \sigma_0[J(t) - J(t - t_r)]$$
$$+ \sigma_p[J(t - t_p) - J(t - t_p - \Delta t_p)] \qquad (5)$$

In order to demonstrate that the effect of the load, σ_p, on the recovery procedure is negligible, let the material exhibit viscoelastic behaviour of the Voigt type:

$$J(t) = J_0(1 - e^{-t/\tau}) \qquad (6)$$

where τ is retardation time. Equation (5) assumes the form:

$$\varepsilon(t) = \sigma_0 J_0(e^{t_r/\tau} - 1)e^{-t/\tau} + \sigma_p J_0(e^{\Delta t_p/\tau} - 1)e^{-(t - t_p)/\tau} \qquad (7)$$

The value of the first term depends on the value of the quantity $\lambda_0 = e^{t_r/\tau} - 1$, since $t_r \gg \tau$, assumes values $\lambda_0 \gg 1$. The value of the second term depends on the value of the quantity $\lambda_p = e^{\Delta t_p/\tau} - 1$, where $\Delta t_p \ll \tau$ and, therefore, $\lambda_p \to 0$. Hence, we see that the recovery process and, accordingly, the 'locked-in' stress/strain and birefringence fields are not affected by very short-duration loads, irrespective of their intensity.

Now the load σ_p may very well exceed the strength of the polymeric material, corresponding to the loading mode and the respective stress-rate applied. In this case, the material is fractured without any effect at all on the 'locked-in' fields. Since machining procedures are practically based on the application of very high shear loading, applied for very short duration at high stress rates, slicing techniques—the most important device for conventional three-dimensional photoelasticity—can be safely applied. For example, with thermosetting resins, slicing can be obtained by means of saw-discs with 150–600 mm diameter, rotating at 3500–6000 rpm or at 30–50 m s^{-1} speed. With 1 tooth mm^{-1}, the duration of the corresponding load application is of the order of 36–20 μs, a figure very safely applicable with 'locked-in' fields.

However, the validity of the above is subject to a very important condition: that the whole creep and creep-recovery procedure follows one and the same master curve for creep compliance and birefringence, respectively, which in turn corresponds to a specific reference temperature. This temperature must never be exceeded, for example during machining, otherwise transition to another master curve with different reference temperature and faster rate of recovery will occur and the 'locked-in' fields will most likely be distorted. Sufficient cooling during machining is, therefore, a necessity.

3. MATERIALS

Photoelastic materials suitable for three-dimensional applications must possess the following properties:

(a) They must be castable at large sizes, a possibility which can be obtained when the material exhibits: (i) low exotherm; (ii) low shrinkage and thermal expansion coefficients; (iii) sufficiently long 'pot life' (duration of the liquid phase) if possible at low viscosity, so that impurities can precipitate and air bubbles, etc., be driven out.

(b) Optical and mechanical homogeneity.

(c) High elastic moduli and optical sensitivity, as well as linear mechanical and optical behaviour.

(d) Low cost of casting preparation, which includes short curing times and a rubbery state, corresponding to low temperatures but, at the same time, to long t_r times in order to ensure long term stability of the 'locked-in' fields. The possibility of producing castings 'on shape' is obviously highly desirable.

(e) The materials should be free from parasitic effects, such as moisture absorption, causing distortion of the 'locked-in' fringe patterns.

(f) Finally, the possibility of modification of the mechanical and optical properties of the materials by means of suitable additives is very important, since in this way models for composite systems can be produced.

Out of many materials used for three-dimensional photoelastic applications,[6,11] epoxy resins[12] appear to conform best with these requirements. Some typical properties of thermosetting resins are presented in Table 1 for comparison.

Epoxy resins are generally produced by condensation of epichlorohydrin and polyhydric phenol. Curing can be effected mainly by means of the various amines or acid anhydrides. Amine-cured epoxy systems are termed *cold-setting*, since they can cure at room temperature, while acid-anhydride ones are *hot-setting*, since they can only cure at elevated temperatures. A large number of modifiers, such as plasticisers, accelerators, reactive or non-reactive diluents, etc., are available, in order to control the properties of the cured polymers, but also the polymerisation reaction, especially with systems exhibiting high exotherm.

Polysulphides, such as thiokols, are excellent plasticisers for cold-setting systems. With hot-setting systems, the number of plasticisers is limited by the high curing temperature and, depending on the particular conditions, reactive or non-reactive diluents can be used as plasticisers. These are mainly used to reduce viscosity and prolong pot-life but they also present plasticisation action by reducing crosslinking density. High proportions would be necessary to produce truly flexible compounds. With increasing curing temperature, they also show a tendency to be driven-off from the surface layers of the casting, sometimes leading to thermal microcracking.[13,14]

TABLE 1
Typical Properties of Some Thermosetting Resins

Property	Phenolics	Polyesters	Epoxies
Volume shrinkage (%)	8–10	4–6	1–2
Tensile modulus (GN m^{-2})	3·11	2·08–4·43	3·11–4·14
Coefficient of thermal expansion ($°C \times 10^6$)	60–80	80–100	60–65

Cold-setting systems are not suitable for three-dimensional applications for the following reasons:

(a) Due to high exotherm the polymerisation process cannot be easily controlled, especially with large castings. Sealed moulds with water-cooled systems are necessary, although they are not always effective due to the low thermal conductivity of the material.

(b) During curing, but also during thermal processing, large thermal stresses develop and the casting is subject to cracking.

(c) During curing at room temperature, i.e. at the initial curing stage, long linear molecular chains are formed while crosslinking occurs at elevated temperatures during a post-curing stage. Stress/strain/birefringence fields induced during the initial stage remain permanently cured in the material. Such parasitic fields cannot be eliminated by annealing if the material must be post-cured in the mould, as could be the case with models cast 'on shape'.

Hot-setting systems exhibit none of these drawbacks, while, in addition, by possessing a long pot-life under conditions of low viscosity, they produce perfectly homogeneous castings, which are free of air bubbles or impurities, which rapidly precipitate, and are of ideal transparency.

A typical example is DGEBA resin (diglycidyl ether of Bisphenol A) as a prepolymer, phthalic anhydride (PA) at 40 phr as curing agent and a non-reactive diluent such as dibutyl phthalate (DBP) as plasticising agent. The curing agent, in the form of small platelets, dissolves into the prepolymer at 80°C, and curing is effected at 100–110°C for a few hours. Thin metal sheet must be used for the moulds, which need to be sufficiently covered with a mould release agent such as a silicone oil. Master curves for creep compliance for such a system are presented in Fig. 1.

3.1. Determination of Time-dependent Behaviour

As stated, the time-dependent behaviour of photoelastic materials over a wide time or frequency range can be described by the corresponding master curves, which are constructed on the basis of the *time–* or *frequency–temperature superposition principle* or *method of reduced*

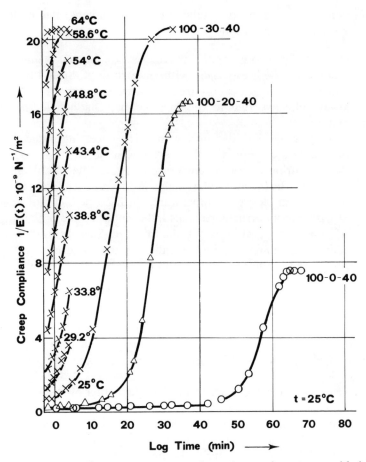

Fig. 1. Creep-compliance master curves for a hot-setting system with individual curves for 30% plasticisation (triads of numbers denote amounts of prepolymer, plasticiser and curing agent).

variables,[10,15] described by the equations

$$\frac{T_0}{T} E_{r,t}(kt) = E_{r,T_0}(t) \tag{8a}$$

for transient loading, and

$$\frac{T_0}{T} E_{r,T}(\omega) = E_{r,T_0}(k\omega) \tag{8b}$$

for harmonic loading, where t is time, ω is radian frequency, $E_{r,T}$ is a characteristic quantity, e.g. a relaxation function or dynamic modulus at temperature T, and k is a *time factor* (for eqn (8a)) or *frequency factor* (for eqn (8b)), expressing the amount of shifting of individual curves corresponding to short time or frequency intervals at various temperatures, along the time or frequency scale, in order to produce a master curve.

By means of the interrelations between viscoelastic spectra,[10] from a given viscoelastic function one can derive another function, for example on the basis of the equation:[16,17]

$$E_r(t) = E'(\omega) - 0 \cdot 40 G'(0 \cdot 40\omega) + 0 \cdot 0146 G''(10\omega) \qquad (9)$$

by which the relaxation modulus $E_r(t)$ can be derived from the dynamic functions corresponding to harmonic loading. Here, $E'(\omega)$ is the extensional storage modulus and $G'(\omega)$ and $G''(\omega)$ are the shear storage and loss moduli, respectively, at frequency ω. Results from

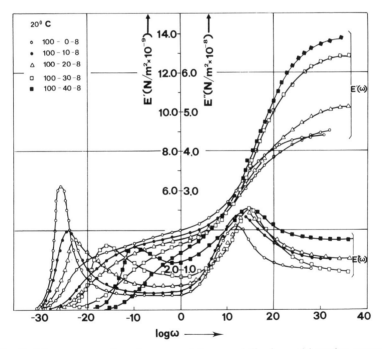

Fig. 2. Master curves for storage and loss moduli of a cold-setting system, DGEBA–Thiokol–TETA).

actual relaxation tests and from conversion through eqn (9) agree quite satisfactorily.[18] For example, with a cold-setting system (a DGEBA resin, cured with 8 phr, i.e. slightly lower than stoichiometric amounts of triethylene tetramine, and plasticised with various amounts of Thiokol LP-3 polysulphide), master curves for storage and loss moduli at 20°C are presented in Fig. 2, while the corresponding relaxation modulus master curves are presented in Fig. 3. The above values have been obtained by means of individual curves constructed from values corresponding to four discrete low frequencies in a temperature range −150–150°C.[18] This wide temperature range allows for the construction of a master curve extending over many logarithmic decades of frequency, exhibiting also an α-transition zone, occurring at very low temperatures. It is important to note that, by use of the proper accessories on the same instrument, both extension and shear dynamic moduli can be determined,[19] to make eqn (9) applicable. The derivation of corresponding dynamic stress or strain optical coefficients presents only minor experimental difficulties.

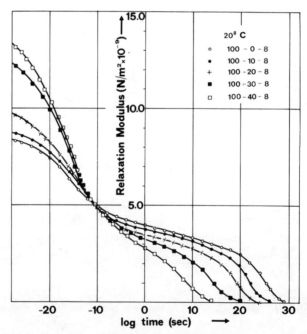

Fig. 3. Master curves for relaxation modulus of the system of Fig. 2, derived by means of eqn (9).

TABLE 2
Comparison of t_r-values Obtained by a Contact and a Non-contact
Method for the Cold-setting System of Fig. 2

Material	t_r (years)	
	From refs 19–22	*From ref. 18*
100-0-8	1·995 (728·17 days)	$3·17 \times 10^{22}$
100-20-8	0·249 (91·02 days)	$3·17 \times 10^{15}$
100-30-8	0·0312 (11·38 days)	$3·17 \times 10^{12}$
100-40-8	0·020 (8·04 days!)	$3·17 \times 10^{6}$

This method, the derivation of the transient functions from the dynamic ones, is perhaps the most reliable, since it is based on a non-contact procedure. This is of cardinal importance, particularly at high temperatures, where the material is very soft and any contact method would provide highly unreliable results.

A striking example is given in Table 2, where the results presented in Fig. 3 are compared with others, obtained by means of a mechanical device attached to the specimen, without any precaution taken to counterbalance its own weight.[19–21] The results are obviously highly unrealistic.

A further example refers to an unplasticised hot-setting polymer (DGEBA/PA) for which a time $t_r = 0·0312$ years is determined by the said method,[22] while t_r for the same material according to Fig. 1 is of the order of $2·25 \times 10^{15}$ years. The latter figure was obtained by means of strain-gauges which, although not highly reliable at high temperatures since they cause local reinforcement of the material, still lead to much more realistic results.

4. MACHINING

Moulding three-dimensional photoelastic models 'on shape' is the most convenient and inexpensive method of manufacture. However, it may present several problems: no high dimensional accuracy can be obtained, in any case not less than 0·2–4% and even then by means of very precisely machined moulds—which increases their cost immensely, a cost which is high anyway, since the moulds are not used for mass production of specimens. Moreover, machining operations might be inevitable because of the particular form of the model considered.

Therefore, machining is a very important procedure, worth examining in connection with photoelastic materials for the additional reason that slicing is an important machining operation, requiring high precision and careful control of the respective parameters.

Machining of photoelastic specimens should be able to produce high surface quality, e.g. with low roughness and no surface microcracks and/or burning due to excessive frictional heat, while no excessive wear of the cutting edges of the tools should occur. Good knowledge of the rheological and thermal properties of particular resins under various conditions is necessary in order that all these requirements be fulfilled. However, the safest way is to determine by trial and error the proper machining parameters on each particular occasion. It is important to examine the cutting procedure during machining, along with the forces associated with it and the respective energy dissipation.

The cross-section of an operating cutting tool is presented in Fig. 4(a), where β is the cutting tool angle, α the slope angle of the rake surface of the cutting tool and $\gamma = 90° - (\alpha + \beta)$, is the relief angle. In addition, b is the cutting depth, b' the thickness of the chip produced and γ is the shear strain developing at the cutting speed v. Cutting is effected along the shear plane OA with inclination φ and shear stress along this plane obviously exceeds the shear yield stress of the material at the corresponding shear rate. The cutting tool exerts a force R on the work material (Fig. 4(b)), which can be resolved into the forces F_c and F_t, in the cutting direction and normal to it, respectively. Alternatively it can be resolved for the forces F_f and F_{fn} in the direction of the rake surface of the cutting tool and normal to it. Now energy dissipation, producing heat, occurs (a) along the rake surface by the friction force F_f, and (b) along the shear plane OA by means of the force F_s parallel to that plane, and it is expressed by the following equation:

$$w = v(F_s \gamma \cos \varphi + F_f . r_t) \qquad (10)$$

where w is work per unit time, $\gamma = \tan \varphi + \cot (\varphi - \alpha)$ and $r_t = b/b'$ is the uncut over cut chip thickness ratio. Part of this work is used to increase the temperature of the specimen, and thermal energy $E_T \gtrless w$ should be removed by sufficient coolant flow (e.g. water-soap solution or even compressed air), in order to prevent distortion of the 'frozen-in' photoelastic field.

The quality of machining depends on the kind of chip produced. Optimum cutting conditions are obtained with continuous chip formation, in which case fluctuations in cutting force and surface roughness

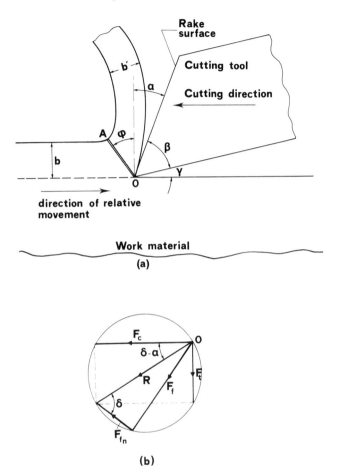

Fig. 4. (a) Geometry of the cutting procedure; (b) forces developing during cutting.

are minimised and good accuracy results. Large elastic chip deformations occur in this case, concerning mainly rubber elastic material or alternatively resins which have undergone plasticisation.

On the contrary, discontinuous chip deformation generally involves brittle fracture and rough surfaces at low machining accuracy are produced.

A 'machinability factor', n_1, increasing with long tool life, good surface finish and low power consumption, is defined as follows:

$$n_1 = \frac{v_m}{v_w p s_f} \tag{11}$$

where v_m is cut volume per unit time, v_w is volume of material worn from tool per unit time, p is net power consumption during cutting, and s_f is cut-surface roughness in microns.

With continuous chip formation, the cut-surface roughness is small, and eqn (11) becomes:

$$n_2 = \frac{v_m}{v_w p} \tag{12}$$

while, with a proper tool material exhibiting minimal wear, it is:

$$n_3 = v_m/p \tag{13}$$

More details on the machining operation, as well as values for the respective parameters pertaining to specific materials can be found in the literature.

5. PARTICULAR PROBLEMS

A number of particular problems arise during the manufacture of three-dimensional photoelastic models, mainly due to their size and these are associated with the inherent properties of the corresponding polymers, such as polymerisation exotherm and shrinkage and their tendency to crack at high temperature rates or in the presence of chemical instability.

With hot-setting systems the exothermic reaction may be of the order of 20–30°C over the cure temperature at the most, even for specimens exceeding 20 kg, i.e. in general no important problems can develop as is the case with cold-setting systems.[25] However, heating the resin/curing agent mixture at high temperatures causes the reactivity of the latter to increase—a factor which should remain under control.

Shrinkage stresses develop during polymerisation in constrained systems such as a casting in a mould or a composite system and they may assume considerably high values. For example, with TETA-cured DGEBA resin containing a rigid spherical inclusion, shrinkage stresses as high as 25·96 MN m^{-2} (264·59 kg cm^{-2}) may develop.[26]

Three-dimensional photoelasticity is a most effective means of investigating shrinkage stresses of composite systems, as presented in Fig. 5,

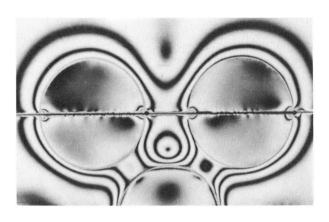

Fig. 6. Dark-field isochromatics of a diagonal slice of the system of Fig. 5.

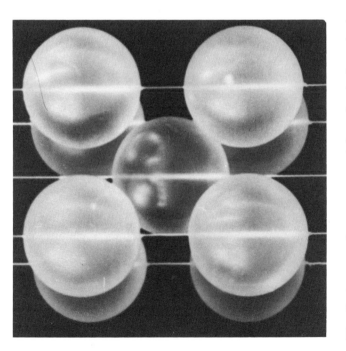

Fig. 5. A composite system consisting of a body-centred cubic arrangement of unplasticised spheres in a plasticised matrix.

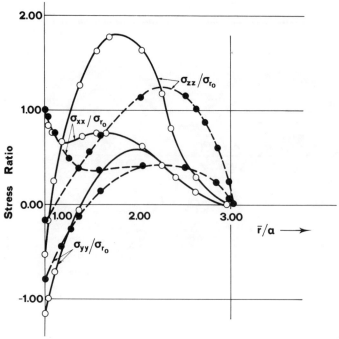

Fig. 7. Distribution of shrinkage stresses along the axis of symmetry to the slice of Fig. 6 (Inclusion-to-matrix moduli ratio (○) $E_2/E_1 = 1.6$; (●) $E_2/E_1 = 1.0$; σ_{r_0}, shrinkage stress at inclusion surface.)

Fig. 8. Surface cracking due to shrinkage stresses in a hot-setting, plasticised composite system.

where a body-centred cubic arrangement of spheres manufactured from an unplasticised hot-setting resin is embedded in a plasticised matrix. Figure 6 shows dark-field isochromatics on a diagonal slice and Fig. 7 the corresponding distribution of shrinkage stresses. Shrinkage stresses, along with stresses developing during thermal processing of the model due to differential thermal contraction, may lead to cracking of the material and the final destruction of the model. An interesting example is associated with the chemical instability of the material at elevated temperatures. A DGEBA/DBP/PA system develops microcracks at temperatures over 120°C, a phenomenon due to driving off DBP from the system. Surface cracking can then be produced by shrinkage stresses (Fig. 8) or even by excessive surface roughness (Fig. 9).

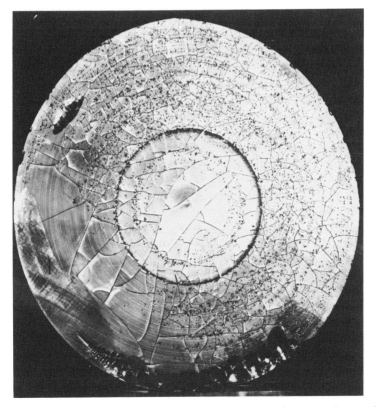

Fig. 9. Surface cracking of a high surface roughness specimen of a hot-setting plasticised epoxy system.

6. EVALUATION OF POTENTIAL AND COST EFFECTIVENESS

Despite the relatively costly procedures associated with stress-freezing and slicing techniques, the method has never ceased to appear in the literature, mainly with problems which cannot be treated efficiently by other experimental methods. The investigation of residual stresses in glass and transparent plastics,[29] of stress concentration in notched structural components,[30] of contact stresses in shrink-fits,[31] the photoplastic simulation[32,33] and the state of stress in orthotropic composites[34] are typical examples. Less costly procedures can be obtained by applying moulding 'on shape',[6,35] while combination with optical techniques such as oblique incidence[36] or holography[37] not only facilitate photoelastic analysis but also extend the possibilities of the method. Concerning applications for design purposes,[38] the method is obviously cost-effective with prototypes constructed for development purposes.

In conclusion, conventional three-dimensional photoelasticity is a method requiring substantial experience, good quality equipment and skilled personnel, while the individual procedures are generally costly, from both the points of view of necessary materials and labour. Therefore, the method is cost-effective only if the value of the information to be derived from the experimental investigation is accordingly high. In general, this is the case with: (a) prototypes used for development purposes; (b) problems which by their nature render all other numerical or experimental methods inapplicable. As a final remark, conventional three-dimensional photoelasticity is a powerful tool for experimental stress analysis, which still retains its unique value for the solution of important technological problems.

REFERENCES

1. Coker, E. G. and Filon, L. N., *A Treatise on Photoelasticity*, Cambridge University Press, 1931.
2. Kuske, A., Das Kunstharz Phenolformaldehyd in der Spannungsoptik, *Forsch. Ing. Wes.*, **9** (1938), 139.
3. Opel, G., Polarisationsoptische Untersuchung räumlicher Spannungs und Dehnungszustände, *Forsch. Ing. Wes.*, **7** (1936), 240.
4. Hetenyi, M., The fundamentals of three-dimensional photoelasticity, *J. Appl. Mech.*, (1938), 149.
5. Frocht, M. M., *Photoelasticity*, Vol. 2, John Wiley and Sons, New York, 1948.

6. Cernosek, J., Three-dimensional photoelasticity by stress freezing, *Exp. Mech.*, **20** (1980), 417.
7. Durelli, A. J. and Lake, P. L., Three-dimensional photoelasticity, *Machine Design* (December 1950), 1.
8. Schwaighofer, J., Extended frozen stress method, *Proc. ASCE, Eng. Mech. Div.*, **EM6** (1962), 1.
9. Frocht, M. M. and Guernsey, R., Further work on the general three-dimensional photoelastic problem, *J. Appl. Mech., Trans. ASME*, E**22** (1955), 183.
10. Ferry, J., *Viscoelastic Properties of Polymers*, John Wiley and Sons, New York, 1961.
11. Dally, J. W. and Riley, W. F., *Experimental Stress Analysis*, McGraw Hill, New York, 1967.
12. Theocaris, P. S., The viscoelastic behavior of epoxy resins in the transition region, NSF-G8188/3, May 1960, Brown University, Rhode Island.
13. Theocaris, P. S. and Paipetis, S. A., Thermal crazing of hot-setting, plasticised epoxy polymers, *Fib. Sci. Technol.*, **7** (1974), 33.
14. Theocaris, P. S., Paipetis, S. A. and Tsangaris, J. M., Thermal crazing phenomena in epoxy resins, *Polymer* **15** (1974), 441.
15. Andrews, R. D. and Tobolsky, A. V., Systems manifesting superposed elastic and viscous behaviour, *J. Chem. Phys.*, **13** (1945), 3.
16. Schwartzl, F. R., Numerical calculation of storage and loss modulus from stress relaxation data for linear viscoelastic materials, *Rheol. Acta*, **10** (1971), 165.
17. Schwartzl, F. R., Numerical calculation of stress relaxation modulus from dynamic data for linear viscoelastic materials, *Rheol. Acta*, **14** (1975), 581.
18. Paipetis, S. A., Theocaris, P. S. and Marchese, A., The dynamic properties of plasticised epoxies over a wide frequency range, *Coll. Polym. Sci.*, **257** (1979), 478.
19. Theocaris, P. S., Viscoelastic properties of epoxy resins derived from creep and relaxation tests at different temperatures, *Rheol. Acta*, **2** (1962), 92.
20. Theocaris, P. S., Rheologic behaviour of epoxy resins in the transition region, *J. Appl. Polym. Sci.*, **8** (1964), 399.
21. Theocaris, P. S. and Hadjijoseph, C., Viscoelastic behaviour of plasticised epoxy polymers in their transition region, *Proc. 4th Int. Congr. Rheol.*, Vol. 3, Interscience Publishers, New York, 1965, p. 485.
22. Theocaris, P. S., Phenomenological analysis of mechanical and optical behaviour of rheo-optically simple materials, In: *The Photoelastic Effect and its Applications*, ed. Jean Kestens, Springer-Verlag, Berlin, 1975, p. 146.
23. Alexander, J. M. and Brewer, R. C., *Manufacturing Properties of Materials*, Van Nostrand Reinhold, London, 1971.
24. Kobayashi, A., Machinability, In: *Encyclopedia of Polymer Science and Technology*, Vol. 8, John Wiley and Sons, New York, 1968, p. 339.
25. Lee, H. and Neville, K., *Handbook of Epoxy Resins*, McGraw-Hill, New York, 1967.
26. Paipetis, S. A., The mechanical behaviour of particle composites in the presence of shrinkage stresses, *Coll. Polym. Sci.*, **257** (1979), 934.

27. Theocaris, P. S. and Paipetis, S. A., Shrinkage stresses in three-dimensional two-phase systems, *J. Strain Analysis*, **8** (1974), 33.
28. Paipetis, S. A., Experimental modelling of composites, In: *Developments in Composite Materials—2*, ed. G. S. Holister, Applied Science Publishers Ltd., London, 1981, p. 39.
29. Render, A. S. and Nickola, W. E., Measurement of residual strains and stresses in transparent materials, *Exp. Techniques*, **8** (1984), 29.
30. Rubayi, N. A. and Taft, M. E., Photoelastic study of axially loaded thick-notched bars, *Exp. Mech.*, **22** (1982), 377. (See also discussion by Rowland Richards, *Exp. Mech.* **24** (1984), 87.)
31. Scurria, N. V. and Doyle, J. F., Photoelastic analysis of contact stresses in the presence of machining irregularities, *Exp. Mech.* **22** (1982), 342.
32. Gomide, H. A. and Burger, C. P., Three-dimensional strain distributions in upset rings by photoplastic simulation, *Exp. Mech.* **21** (1981), 361.
33. Burger, C. P. and Gomide, H. A., Three-dimensional strains in rolled slabs by photoplastic simulation, *Exp. Mech.*, **22** (1982), 441.
34. Chandrashekhare, K., Abraham Jacob, K. and Prabhakaran, R., Towards stress freezing in birefringent orthotropic composite models, *Exp. Mech.*, **17** (1977), 317.
35. Cernosek, J., Mold for casting photoelastic models, *Exp. Techniques*, **8** (1984), 21.
36. Allison, I. M., Application of the oblique-incidence technique for economic photoelastic analysis, *Exp. Techniques*, **8** (1984), 25.
37. Dai Fu-lung and Chung Kuo-Cheng, Use of the holophotoelastic method for three-dimensional stress analysis, *Exp. Mech.* **22** (1982), 468.
38. Narayanan, T., Narayanan, R. and Sambandam, T., Photoelastic investigation of a channel cover of bleed condenser, *Exp. Mech.*, **20** (1980), 309.

INDEX